T0143600

RESEARCH PRACTITIONER'S HANDBOOK ON BIG DATA ANALYTICS

RESEARCH PRACTITIONER'S HANDBOOK ON BIG DATA ANALYTICS

S. Sasikala, PhD
Renuka Devi D, PhD

Raghvendra Kumar, PhD
Editor

First edition published 2023

Apple Academic Press Inc.
1265 Goldenrod Circle, NE,
Palm Bay, FL 32905 USA
760 Laurentian Drive, Unit 19,
Burlington, ON L7N 0A4, CANADA

CRC Press
6000 Broken Sound Parkway NW,
Suite 300, Boca Raton, FL 33487-2742 USA
4 Park Square, Milton Park,
Abingdon, Oxon, OX14 4RN UK

© 2023 by Apple Academic Press, Inc.

Apple Academic Press exclusively co-publishes with CRC Press, an imprint of Taylor & Francis Group, LLC

Library and Archives Canada Cataloguing in Publication

Title: Research practitioner's handbook on big data analytics / S. Sasikala, PhD, D. Renuka Devi, Raghvendra Kumar, PhD.
Names: Sasikala, S., 1970- author. | Devi, D. Renuka, 1980- author. | Kumar, Raghvendra, 1987- author.
Description: First edition. | Includes bibliographical references and index.
Identifiers: Canadiana (print) 20220437475 | Canadiana (ebook) 20220437491 | ISBN 9781774910528 (hardcover) | ISBN 9781774910535 (softcover) | ISBN 9781003284543 (ebook)
Subjects: LCSH: Big data—Handbooks, manuals, etc. | LCSH: Big data—Research—Handbooks, manuals, etc. | LCSH: Data mining—Handbooks, manuals, etc. | LCSH: Electronic data processing—Handbooks, manuals, etc. | LCGFT: Handbooks and manuals.
Classification: LCC QA76.9.B45 S27 2023 | DDC 005.7—dc23

Library of Congress Cataloging-in-Publication Data

....

CIP data on file with US Library of Congress

....

ISBN: 978-1-77491-052-8 (hbk)
ISBN: 978-1-77491-053-5 (pbk)
ISBN: 978-1-00328-454-3 (ebk)

About the Authors

S. Sasikala, PhD

S.Sasikala, PhD, is Associate Professor and Research Supervisor in the Department of Computer Science, IDE, and Director of Network Operation and Edusat Programs at the University of Madras, Chennai, India. She has 23 years of teaching experience and has coordinated computer-related courses with dedication and sincerity. She has acted as Head-in-charge of the Centre for Web-based Learning for three years, beginning in 2019. She holds various posts at the university, including Nodal Officer for the UGC Student Redressal Committee, Coordinator for Online Course Development at IDE, President for Alumni Association at IDE. She has been an active chair in various Board of Studies meetings held at the institution and has acted as an advisor for research. She has participated in administrative activities and shows her enthusiastic participation in research activities by guiding research scholars, writing and editing textbooks, and publishing articles in many reputed journals consistently. Her research interests include image, data mining, machine learning, networks, big data, and AI. She has published two books in the domain of computer science and published over 27 research articles in leading journals and conference proceedings as well as four book chapters, including in publications from IEEE, Scopus, Elsevier, Springer, and Web of Science. She has also received best paper awards and women's achievement awards. She is an active reviewer and editorial member for international journals and conferences. She has been invited for talks on various emerging topics and chaired sessions in international conferences.

Renuka Devi D, PhD

Renuka Devi D, PhD, is Assistant Professor in the Department of Computer Science, Stella Maris College (Autonomous), Chennai, India. She has 12 years of teaching experience. Her research interests include data mining, machine learning, big data, and AI. She actively participates in continued learning through conferences and professional research. She has published eight research papers and a book chapter in publications from IEEE, Scopus, and Web of Science. She has also presented papers at international conferences and received best paper awards.

About the Editor

Raghvendra Kumar, PhD

Raghvendra Kumar, PhD, is Associate Professor in the Computer Science and Engineering Department at GIET University, India. He was formerly associated with the Lakshmi Narain College of Technology, Jabalpur, Madhya Pradesh, India. He also serves as Director of the IT and Data Science Department at the Vietnam Center of Research in Economics, Management, Environment, Hanoi, Viet Nam. Dr. Kumar serves as Editor of the book series Internet of Everything: Security and Privacy Paradigm (CRC Press/Taylor & Francis Group) and the book series Biomedical Engineering: Techniques and Applications (Apple Academic Press). He has published a number of research papers in international journals and conferences. He has served in many roles for international and national conferences, including organizing chair, volume editor, volume editor, keynote speaker, session chair or co-chair, publicity chair, publication chair, advisory board member, and technical program committee member. He has also served as a guest editor for many special issues of reputed journals. He authored and edited over 20 computer science books in field of Internet of Things, data mining, biomedical engineering, big data, robotics, graph theory, and Turing machines. He is the Managing Editor of the *International Journal of Machine Learning and Networked Collaborative Engineering*. He received a best paper award at the IEEE Conference 2013 and Young Achiever Award–2016 by the IEAE Association for his research work in the field of distributed database. His research areas are computer networks, data mining, cloud computing and secure multiparty computations, theory of computer science and design of algorithms.

Contents

Abbreviations

ABC	Ant Bee Colony
ABC	Artificial Bee Colony
ACO	Ant Colony Optimization
AI	Artificial Intelligence
ANNs	Artificial Neural Networks
APIs	Application Programming Interfaces
BA	Bat Algorithm
BI	Business Intelligence
BMU	Best Matching Unit
CDC	Centers for Disease Control and Prevention
CHOA	Children's Healthcare of Atlanta
CNNs	Convolutional Neural Networks
CT	Computed Tomography
DA	Dragonfly Algorithm
DBNs	Deep Belief Networks
DL	Deep Learning
EMRs	Electronic Medical Records
ETL	Extract–Transform–Load
FHCN	Family HealthCare Network
FS	Feature Selection
GA	Genetic Algorithms
GANs	Generative Adversarial Networks
GWO	Grey Wolf Optimization Algorithm
HDFS	Hadoop Distributed File System
HQL	Hive Query Language

JSON	JavaScript Object Notation
KNN	K-Nearest Neighbor
KPIs	Key Performance Indicators
LR	Linear Regression
LSTMs	Long Short-Term Memory networks
MDP	Markov Decision Process
ML	Machine Learning
MLP	Multilayer Perceptron
NPCR	National Program of Cancer Registries
OFS	Online Feature Selection
OLAP	Online Analytic Processing
PSO	Particle Swarm Optimization
RBFNs	Radial Basis Function Networks
RBMs	Restricted Boltzmann Machines
ReLU	Rectified Linear Unit
RNNs	Recurrent Neural Networks
SI	Swarm Intelligence
SOMs	Self-Organizing Maps
SVM	Support Vector Machine
WOA	Whale Optimization Algorithm

Preface

Throughout the learning behavior among academicians, researchers, and students around the globe, we observe unprecedented interest in big data analytics. As decades pass by, big data analytics knowledge transfer groups have been intensively working on shaping various nuances and techniques and delivering them across the country. This would have not been possible without the Midas touch of researchers who have been extensively carrying out research across domains and connecting big data analytics as a part of other evolving technologies. This book discusses major contributions and perspectives in terms of research over big data and how these concepts serve global markets (IT industry) to lay concrete foundations on the same technology. We hope that this book will offer a wider connotation to researchers and academicians in all walks and on par with their ways of using big data analytics as a theoretical and practical style.

Introduction

With the recent developments in the digital era, data escalate at a rapid rate. Big data refers to an assortment of data that are outsized and intricate so that conventional database administration systems and data processing tools cannot process them. This book mainly focuses on the core concepts of big data analytics, tools, techniques, and methodologies from the research perspectives. Both theoretical and practical approaches are handled in this book that can cover a broad spectrum of readers. This book would be a complete and comprehensive handbook in the research domain of big data analytics.

Chapter 1 briefs about the fundamentals of big data, terminologies, types of analytics, and big data tools and techniques.

Chapter 2 outlines the need for preprocessing data and various methods in handling the same. Both text and image preprocessing methods are also highlighted. In addition to that, challenges of streaming data processing are also discussed.

Chapter 3 briefs on various featured selection methods and algorithms, and research problems related to each category are discussed with specific examples.

Chapter 4 describes the core methods of big data streams and the prerequisite for parallelization. This chapter also enlightens on the streaming architecture. Hadoop architecture is comprehensively mentioned with the components of parallel processing.

Chapter 5 updates on the big data classification techniques, and various learning methodologies are explained with examples. To extend the same, deep learning algorithms and architectures are also briefed.

Chapter 6 highlights application across verticals with research problems and solutions.

CHAPTER 1

Introduction to Big Data Analytics

ABSTRACT

This chapter briefs about fundamentals of big data analytics, terminologies, types of data analytics, big data tools and techniques. The fundamentals of big data analytics extend into understanding the various types of big data, its 5V's characteristics, and the sources of big data. The core types of big data analytics are explained with examples. Big data analytics refers to a group of tools and techniques that use new methods of incorporation to retrieve vast volumes of unknown knowledge from large datasets that are more complex, large-scale, and distinct from conventional databases. In recent times, various tools have been developed for deeper analytics and visualization. The commonly used big data analytics tools are discussed elaborately covering the key features, applications and highlighting the potential advantages.

1.1   INTRODUCTION

The word "big data" applies to the evolution and application of technology that provide people with the right knowledge from a mass of data at the right moment; it has been growing exponentially in its culture for a long time. The task is not only to deal with exponentially growing data volumes but also with the complexities of increasingly heterogeneous formats and increasingly dynamic and integrated data management (Anuradha, 2015). Its meaning varies according to the groups that involve a customer or service provider.

Research Practitioner's Handbook on Big Data Analytics. S. Sasikala, PhD, D. Renuka Devi, & Raghvendra Kumar, PhD (Editor)
© 2023 Apple Academic Press, Inc. Co-published with CRC Press (Taylor & Francis)

Big data, invented by the network giants, describes itself as a solution intended to allow anyone access to giant datasets in real time. It is difficult to exactly describe big data because the very notion of big differs from one field to another in terms of data volume. It does not describe a collection of technologies but describes a group of techniques and technologies.

This is an evolving field, and the meaning shifts as we understand how to apply this new concept and leverage its value. Digital data generated is partly the product of the use of Internet-connected devices. Smartphones, laptops, and computers, therefore, relay information about their customers. Linked smart objects relay knowledge about the use of ordinary objects by the user.

In addition to connected computers, information comes from a broad variety of sites: population data, climate data, science and medical data, data on energy use, and so on. Data include information on the location, transport, preferences, usage patterns, recreational activities, and ventures of users of smartphones, and so on. However, information about how infrastructure, machinery, and facilities are used is also available (Loshin, 2013). The amount of digital data is rising exponentially with the ever-increasing number of Internet and cell phone users.

In the last decade, the amount of data that one must contend with has risen to unprecedented heights and, at the same time, the price of data storage has systematically decreased. Private firms and academic organizations collect terabytes of data from devices such as cell phones and cars regarding the experiences of their customers, industry, social media, and sensors. This era's challenge is to make sense of this sea of knowledge.

This is where the study of big data comes into focus. Big data analytics primarily means gathering data from multiple sites, mixing it in a manner where researchers can consume it, and eventually providing data items that are beneficial to the business of the enterprise. The essence of big data analytics is the method of transforming vast volumes of unstructured raw data retrieved from multiple sites to a data product usable for organizations.

It is important to use this knowledge for predictive forecasting, as well as for marketing and also involving many other uses. It would not be possible to execute this task within the specified time frame if it uses the conventional solution, since the storage and computing space will not be suitable for these types of tasks. To explain the meaning of big data, it has a clearer description. The more reliable version is given by Kim and Pries (2015):

Data that is vast in scale concerning the retrieval system, with several organized and unstructured data to be processed comprising multiple data patterns. Data is registered, processed, and analyzed from traffic flows and downloading of music, to web history and medical information, to allow infrastructure and utilities to deliver the measurable performance that the world depends on every day. If it just keeps hanging on to the information without analyzing it, or if it does not store the information, finding it to be of little use, it could be to the detriment of the organization.

These businesses manage all the things that they do on their website and use them to create income (Loshin, 2013; Anuradha, 2015) for overall improved customer experience, as well as for their benefit. There are several examples of these types of activities that are available and they are rising as more and more enterprises understand the power of information. For technology researchers, this poses a challenge to come up with more comprehensive and practical solutions that can address current problems and demands.

The information society is now transitioning to a knowledge-based society. It needs a greater volume of data to extract better information. The knowledge society is a society in which data plays a significant role in the economic, cultural, and political arenas. Thus, big data analytics play a significant role in all facets of life.

1.2 WIDER VARIETY OF DATA

The range of sites of data continues to grow. Internally based operating systems, such as enterprise resource planning and CRM applications, have traditionally been the main source of data used in the predictive analysis (Kim and Pries, 2015). However, the complexity of data sizes that feed into the empirical processes is increasing to expand information and understanding and to provide a broader spectrum of data sizes such as:

- Data on the Internet (i.e., clickstream, social networking, connections to social networks).
- Main (i.e., polls, studies, observations) study.
- Secondary analysis (i.e., demand and competitive 1data, business surveys, customer data, organization data).
- Data on positions (i.e., data on smart devices, geospatial data).
- Picture evidence (i.e., film, monitoring, satellite images).

- Data on supply chain (i.e., electronic data interchange (EDI), distributor catalogs and prices, consistency data).
- Information about devices (i.e., sensors, programmable logic controllers (PLCs), radio frequency (RF) devices, Laboratory Information Management System (LIMs), telemetry).

The broad spectrum of data adds to problems in ingesting the data into data storage. The variation of details often complicates the transformation (or the transformation of data into a shape that can be utilized in the processing of analytics) and the computational calculation of data processing.

Big data is considered a series of vast and dynamic datasets that are challenging to store and process utilizing conventional databases and data analysis techniques. Big data is obtained from both conventional and modern channels that can be used for study and review when accurately refined. Over time, organizations are rising and are often increasing rapidly with this knowledge produced by these organizations.

The challenge is to provide a website that includes the entire data with a single, coherent view. Another task is to organize this knowledge in detail so that it makes sense and is beneficial. Big data is constantly generated from all around us. The processing of such an immense volume of data is the duty of social networking platforms and digital outlets. Sensors, mobiles, and networks are the key to how this massive volume of data is transmitted (Loshin, 2013).

1.3 TYPES AND SOURCES OF BIG DATA

1.3.1 TYPES OF DATA

The overview of categories of data is presented in this section.

1.3.1.1 STRUCTURED DATA

Data that is ordered in a predefined schema and has a fixed format is considered structured data (Figure 1.1). Examples of organized data provide data from conventional databases and archives such as mainframes, SQL Server, Oracle, DB2, Sybase, Access, Excel, txt, and Teradata. The method of relational database management deals with mostly this kind of knowledge.

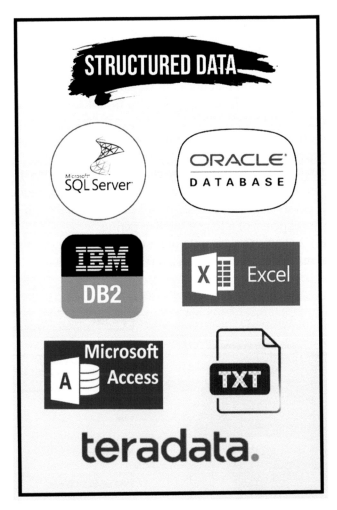

FIGURE 1.1   Structured data.

1.3.1.2   MULTISTRUCTURED DATA

Multistructure data is unmodeled, it must be ordered, so it is overlooked even if there might be a schema. It can be learned from human and computer experiences, which includes data on developing economies, e-commerce, and other third-party data, such as temperature, currency conversion, demographics, tables, etc.

1.3.1.3   METADATA

Metadata is information that does not represent data but contains knowledge regarding the attributes and structure of a dataset. Metadata monitoring is important to the collection, storing, and interpretation of big data because it offers knowledge regarding the birth of data, as well as all its collection measures (Figure 1.2).

Metadata allows managing data in certain situations. It can, for instance, maintain certain metadata regarding the resolution of the picture and the number of colors used. Of course, it will get this detail from the graphic image, at the expense of a longer loading period.

Figure 1.2 shows the example of metadata and contains the details of the image, resolution, and other characteristics of the image.

DATA   **META DATA**

FIGURE 1.2   Metadata.

1.3.1.4 UNSTRUCTURED DATA

Unstructured data is considered unstructured data (Figure 1.3) and it is not possible to view such data using conventional databases or data structures. Social networking info pertains to these, such as chatter, text analytics, blogs, messages, mentions, clicks, marks, etc.

The relationship between structured data (data that is easy to describe, index, and analyze) and unstructured data (data that tends to avoid easy classification, requires a lot of storage space, and is typically more difficult to analyze) is "level set" (Engelbrecht and Preez, 2020).

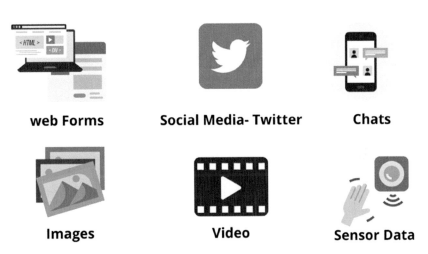

FIGURE 1.3   Unstructured data.

Unstructured data is content that lacks a predetermined data model or does not fit well into a relational database. Text-heavy unstructured data is common, but it may also contain information like hours, numbers, and statistics.

The word semiorganized data is used to define structured knowledge that does not fall within a formal data model framework. However, tags that isolate textual elements include semistructured data, which requires the potential to implement hierarchies within the data.

- Every 2 years, the amount of data (all data, everywhere) doubles.
- It is getting more open regarding its planet. As it grows more familiar with parting with data that used to feel sacred and confidential and thus are starting to acknowledge this.
- The latest data were largely unstructured. Unstructured data accounts for almost 95% of the latest data, while structured data accounts for just 5%.
- Unlike structured data, which continues to evolve more linearly, unstructured data tend to grow steadily.

Unstructured material is vastly therefore underused. Imagine large reserves of oil or other natural resizes ready to be exploited that are only just lying there. As of now, that is the real status of unstructured records. Tomorrow would be a different story because there is a lot of profit to be created for clever people and businesses that can effectively exploit unstructured data.

1.3.2   SOURCES OF BIG DATA

The various sources of big data are given in Figure 1.4.

Social media: Large data firms such as Facebook and Google receive data from whatever operations they undertake. Facebook, Twitter, LinkedIn, journals, SlideShare, Instagram, chatter, WordPress, Jive, and so on are other examples.

Public web: This includes Wikipedia info, healthcare, the World Bank data, economy, weather, traffic, and so on.

Archives: This covers records with all materials, such as medical records, consumer communications, insurance reports, documents scanned, etc.

Docs: Big data outlets provide documents in all format, like HTML, CSV, PDF, XLS, Word, XML, etc.

Media: Images, videos, audio, live streams, podcasts, etc.

Data storage: The different databases and file structures used to archive data, thus function here in this as a big data outlet.

Machine log information: System data, device logs, audit logs, comprehensive CDR call histories, numerous smartphone applications, mobile positions, etc.

Sensor information: Information from medical device-connected systems, road monitors, rockets, traffic surveillance software, video games, home appliances, air-conditioning equipment, office buildings, etc. Thus, large data is a combination of data that is unstructured, organized, and multistructured.

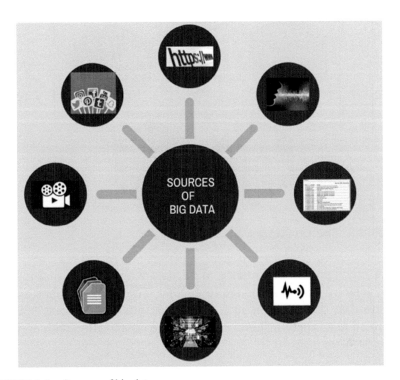

FIGURE 1.4   Sources of big data.

1.4   CHARACTERISTICS OF BIG DATA

Now that have set the groundwork for future debates, let us move on and speak about the first capabilities of big data. It must have more than one attribute, usually referred to as the five Vs in Figure 1.5 (Renuka Devi and Sasikala, 2020) for a dataset to be called big data.

These five attributes of big data are used to help distinguish knowledge classified as big from other data sizes. Doug Laney initially described several of them in the early 2001 when he published an article explaining the effect on enterprise data warehouses of the scale, pace, and variety of e-commerce data.

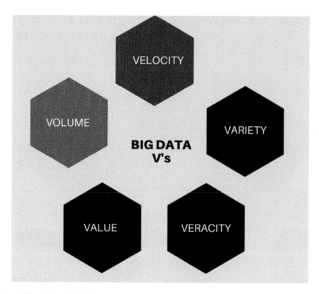

FIGURE 1.5   Five Vs of big data.

To highlight the importance of all data with a low signal-to-noise ratio, veracity has been applied to this list. The use of big data aims to execute data collection in such a way that high-quality outcomes are generated in a timely fashion, supplying the organization with optimum value.

1.4.1   VOLUME

Business statistics in previous years applied only to the details provided by their workers. Now, as the use of technology increases, it is not just employee-generated data but also machine-generated data used by businesses and their customers. Also, people are sharing and uploading too much material, images, photographs, tweets, and so on, with the growth of social networking and other Internet tools. Just imagine, the population of the planet is 7 billion, and nearly 6 billion of them have mobile phones.

There are many sensors in the cell phone itself, such as a gyro-meter, which generates data for each event and is now being collected and analyzed (Saltz et al., 2020). From Table 1.1, numerous machine memory sizes are mentioned to offer an understanding of the conversions between different devices.

TABLE 1.1   Memory Sizes.

1 bit	Binary digit
8 bits	1 byte
1024 bytes	1 KB (kilobyte)
1024 KB	1 MB (megabyte)
1024 MB	1 GB (gigabyte)
1024 GB	1 TB (terabyte)
1,024 TB	1 PB (petabyte)
1024 PB	1 EB (exabyte)
1024 EB	1 ZB (zettabyte)
1024 ZB	1 YB (yottabyte)
1024 YB	1 brontobyte
1024 brontobytes	1 geophyte

In the big data sense, as we speak about volume, it is a large quantity of data concerning the computing method that cannot be obtained, stored, and analyzed using conventional methods. It is data at rest that is already gathered and data that is continuously often produced by streaming.

Take Facebook, for example, they have 2 billion active users who share their status, images, videos, feedback on each other's messages, interests, dislikes, and several more things constantly using this social networking platform. A daily 600 TB of data is absorbed into the Facebook servers, as per the statistics given by Facebook. Figure 1.6 shows the graph below which displays the details that occurred in previous years, the present scenario, and where it will be going in the future (Michael and Miller, 2013).

We take another example of an airplane with a helicopter. For every hit of flight time, one statistic indicates that it produces 10 TB of data. Now think of how the volume of data produced will exceed several petabytes per day, involving thousands of flights per day (Michael and Miller, 2013).

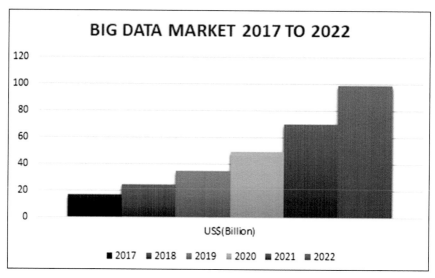

FIGURE 1.6   Data growth.

The volume of data produced in the last 2 years is equivalent to 90% of the data ever made, per 1.2 years, the world's data doubles. One study further states that by 2020, 40 zettabytes of data will be produced. Not that long ago, it was deemed a challenge to produce such a large volume of data as the cost of storage was very high.

But now, when the cost of storage is declining, it is no longer a problem. Solutions such as Hadoop and numerous algorithms that assist in ingesting and analyzing this immense volume of data often render it as look resourceful. Velocity is the second feature of big data. Let us explore what it is.

1.4.2   VELOCITY

Velocity is the pace at which the data is produced, or how rapidly the data arrives. It may term its data in motion in simpler terms. Imagine the sum of data every day received from Twitter, YouTube, or other social networking platforms. They must store it, process it, and be able to recover it somehow later. Here are a few explanations of how data is growing rapidly:

- On each trading day, the New York stock exchange collects 1 TB of data.

- One-hundred and twenty hits of videos are posted per minute to YouTube.
- Data created by modern vehicles; thus approximately 100 sensors are available to track each item from fuel and tire pressure to obstacles around it. Every minute, 200 million emails are sent.
- With the example of developments in social networking, more knowledge indicates more revealing details regarding groups of citizens in various territories.

Figure 1.7 indicates the number of time people spent on common websites for social networking. Based on these user habits, imagine the frequency of data produced. This is merely a snapshot of what is going on out there. The period over which data can make sense and be useful is another component of velocity.

Over time, will it age and reduce the value, or will it remain worth permanently? This is also critical since it can confuse one, if the data ages and loses validity over time, so maybe over time. It addressed two features of big data so far. Variety is the third one. Now let us explore it.

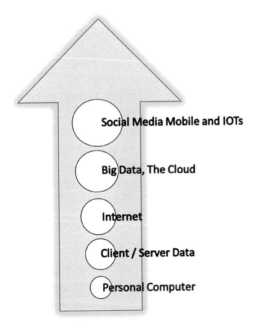

FIGURE 1.7 Velocity of data.

1.4.3   VARIETIES

The review of the classification of data is done in this portion. It may be data that is organized or unstructured. For the knowledge that has a predefined schema or which has a data model rather with predefined columns, data forms, and so on, structured data is chosen, on the other hand, unstructured data has none of these attributes.

Figure 1.8 provides a long list of records, such as papers, addresses, text messages from social media, photographs, still photos, audio, graphs, and sensor performance from all forms of computer-generated records, computers, RFID marks, computer logs, and GPS signals from mobile phones, and more. In distinct chapters, in this novel chapter, we are learning more information regarding structured and unstructured data.

DATA VARITIES

FIGURE 1.8   Data variation.

Let us take one example, 30 billion pieces of content are posted on Facebook every month. There are 400 million messages sent every day. Every month, 4 billion hits of videos are viewed on YouTube. These are both indicators of the production of unstructured data that needs to be processed, either for a stronger customer interface or for the businesses themselves to produce sales. Veracity is the fifth feature of big data.

1.4.4   VERACITY

This particular characteristic deals with knowledge ambiguity. It could be due to low data quality or also because of data noise. It is human behavior to say that knowledge given to us is not always trusted by us. This is one of the main factors that reveal that the evidence they use for decision-making is not trusted by one in three company leaders.

Before analysis and decision-making, it can be considered in a way that velocity and variety are dependent on clean data, whereas veracity is the opposite of these characteristics as it is derived from data uncertainty. The sources of data veracity are presented in Figure 1.9.

Sources - Veracity

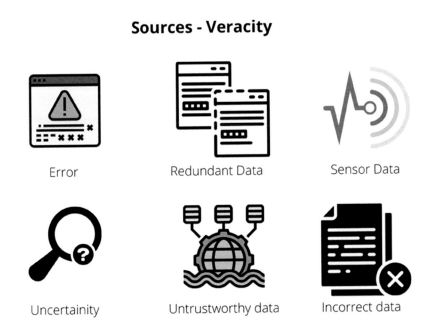

Error	Redundant Data	Sensor Data
Uncertainity	Untrustworthy data	Incorrect data

FIGURE 1.9 Veracity—sources.

To remove confusion, the biggest problem is that they do not have time to clean up streaming data or high-speed data. Machines and sensors produce data such as event data, and if it is hesitant to clean and process it first, the data can lose importance. So, taking account of confusion, it must process it as is.

Veracity is all about misunderstanding and how much confidence one has with data, so it might be that it must redefine trustworthy data with a particular meaning while using it in terms of the sense of big data. It is, in my view, the way it uses data or analyzes it to make decisions. It affects the importance and effect of the decisions it makes because of the faith one has in its data. Now let us look at the fifth big data characteristic, which is uncertainty.

1.4.5 VALUE

In terms of large data, this is the most significant vector, but it is not especially synonymous with big data, and it is equally valid with small data, too. Now it is time to determine if it is worth storing the data and investing in storage, either on-premises or in the cloud, after resolving all the other Vs, length, velocity, variety, variability, and veracity, which requires a lot of time, commitment, and energy.

One aspect of value is that before we can use it to give valuable information in return, we must store a huge amount of data. Earlier, it was lumbered with enormous costs by storing this volume of data, but now storage and recovery technology are so much less costly. One wants to be sure that the data gives value to its organization. To satisfy legal considerations, the study must be done.

1.5 DATA PROPERTY TYPES

1.5.1 QUALITATIVE AND QUANTITATIVE PROPERTIES

There are two main types of properties (Table 1.2):

- Qualitative properties are properties that can be detected, but with a numerical outcome, they cannot be evaluated or calculated (Tsai et al., 2015). To describe a given subject in a more abstract way, including even impressions, opinions, views, and motivations, it uses this type of property.

  This gives a subject breadth of comprehension but also makes it more challenging to examine. It regards this form of property as unstructured. The measurement is nonstatistical if the data type is qualitative.
- Qualitative characteristics ask (or answer) why?
- Numbers and statistical equations rely on quantitative properties and can be measured and computed. It regards this kind of property as organized and statistical. How much is questioned (or answered) by quantitative properties? Yeah, or how many?

Its characteristics can be interpreted as quantitative data on a specific subject. If the characteristic of the property is qualitative, it can be converted into a quantitative one by supplying the numerical details of that characteristic for statistical analysis (Saltz et al., 2020; Michael and Miller, 2013).

TABLE 1.2   Qualitative and Quantitative Properties.

Factor	Qualitative	Quantitative
Meaning	The data in which the classification of objects is based on attributes and properties	The data which can be measured and expressed numerically
Analysis	Nonstatistical	Statistical
Type of data	Unstructured	Structured
Question	Why?	How many or how much?
Used to	Get an initial understanding	Recommends final cause of action
Methodology	Exploratory	Conclusive

1.5.2 *DISCRETE AND CONTINUOUS PROPERTY*

The discrete and continuous properties are given in Table 1.3.

- Discrete is a category of statistical data that can only assume a fixed number of different values and lacks an underlying order often. Often recognized as categorical, since it has divisions that are different, intangible.
- Continuous is a category of statistical knowledge that within a specified range will assume all the possible values. If an infinite and uncountable range of values may be taken from a domain, then the domain is referred to as continuous.

TABLE 1.3   Discrete and Continuous Properties.

Factor	Discrete	Continuous
Meaning	This applies to a vector that considers the number of independent values to be finite	It refers to a vector that considers the number of various values to be infinite and uncountable
Represented by	Lines separated	Points linked
Values (provenance)	By counting, values are collected	Values are gained by measuring
Values (assume)	Distinct or distinct values	For the two values, every meaning
Classification	Nonoverlap	Overlapping over

1.6 BIG DATA ANALYTICS

Big data typically applies to data that extends traditional databases and data mining techniques' usual storage, retrieval, and computational power. Big data requires software and techniques as a resistance that can be used to evaluate and derive trends from large-scale data (Goul et al., 2020). Study of organized data progresses due to the variety and speed of the data that are manipulated.

Therefore, the large range of data suggests that the structures in place would be able to aid in the processing of data and are no longer adequate to interpret data and generate reports. The research consists of efficiently identifying the associations between the data across a spectrum which is constantly shifting the data to aid in the utilization of it.

Big data analytics relates to the method by which vast datasets are gathered, processed, and are evaluated to uncover numerous trends and useful knowledge. Big data analytics refers to a group of tools and techniques that use new methods of incorporation to retrieve vast volumes of unknown knowledge from large datasets that are more complex, large-scale, and distinct from conventional databases. It mainly focuses on overcoming new issues or old problems in the most productive and reliable way possible.

1.6.1 TYPES OF BIG DATA ANALYTICS

Big data is the compilation of large and dynamic databases and data types, including huge numbers of data, social network processing applications for data mining, and real-time data. A vast volume of heterogeneous digital data remains where, in terms of terabytes and petabytes, massive datasets are measured (Saltz et al., 2020). The various types of analytics are discussed below (Figure 1.10).

1.6.1.1 DESCRIPTIVE ANALYTICS

This consists of posing the question: What's going on? (Figure 1.11) It is a preliminary step in the collection of data that provides a historical dataset. Methods of data mining coordinate knowledge and help discover trends that provide insight. Descriptive analytics offers potential patterns and probabilities and provides an understanding of what could happen in the future.

FIGURE 1.10 Types of analytics.

FIGURE 1.11 Descriptive analytics.

1.6.1.2 DIAGNOSTIC ANALYTICS

It consists of answering the question: why did it come about? (Figure 1.12) Diagnostic analytics search for a problem's root cause. To assess whether anything happens, it is included. This form seeks to recognize the origins of incidents and actions and understand them.

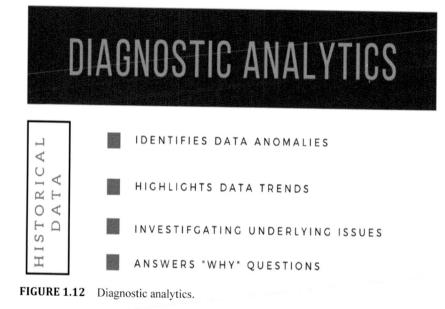

FIGURE 1.12   Diagnostic analytics.

1.6.1.3 PREDICTIVE ANALYTICS

It consists of answering the following question: What is going to happen? (Figure 1.13) To foresee the future, it uses past evidence. Forecasting is all around it. To evaluate existing data and build simulations about what might happen, predictive analytics utilizes multiple tools, such as data mining and artificial intelligence.

1.6.1.4 PRESCRIPTIVE ANALYTICS

It includes posing the question: What is to be done? (Figure 1.14) It is committed to determining the best step to take. Descriptive analytics includes past evidence and what could happen are anticipated through predictive analytics. To identify the right answer, prescriptive analytics utilizes these criteria.

PREDICTIVE ANALYTICS

HISTORICAL DATA

- CREATE DATA MODELS
- FORECAST FUTURE OUTCOMES
- FILL THE GAPS IN AVAILABLE DATA
- ANSWERS "WHAT" MIGHT HAPPEN

FIGURE 1.13 Predictive analytics.

PRESCRIPTIVE ANALYTICS

HISTORICAL DATA

- ESTIMATES OUTCOME BASED ON VARIABLES
- OFFERS SUGGESTIONS ABOUT THE OUTCOMES
- USE AI. ML ALGORITHMS
- ANSWERS "IF.THEN" QUESTIONS

FIGURE 1.14 Prescriptive analytics.

1.6.1.5 DECISIVE ANALYTICS

A set of techniques for visualizing information and recommending courses of action to facilitate human decision-making when presented with a set of alternatives.

1.6.1.6 STREAMING ANALYTICS

Streaming analytics is the real-time analytics, where the data are collected from sensors or devices (Figure 1.15). This kind of analytics help us to identify and understand the pattern of data being generated and can provide immediate analytics and solutions.

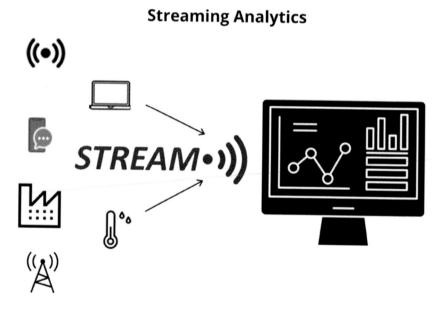

FIGURE 1.15   Streaming analytics.

1.6.1.7 LOCATION ANALYTICS

Location analytics involved much with geographical location analysis (Figure 1.16).

The key features are as follows:

— Enhance business system with location-based predictions with the geographical analysis.
— Maps are used for visual analytics and locate the targets groups.
— Spatial analytics—combining geographical information systems with another type of analytics.
— Explore the temporal and special patterns to locate specific activities or behavior.
— Rich data collection: combines consumer's demographic details, map, lifestyle, and other socio-environmental factors.
— Presence of maps/GPS on all electronic gadgets.

LOCATION ANALYTICS

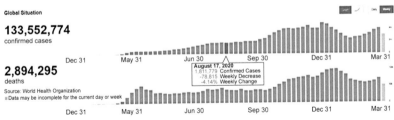

FIGURE 1.16 Location analytics.

Source: World Health Organization covid cases.

1.6.1.8 WEB ANALYTICS

This type of analytics is based on the usage of websites and interactions (Figure 1.17).

The main features are as follows:

- Now: the study of the behavior of web users.
- Future: the study of one mechanism for how society makes decisions.
- Example: behavior of web users:
  - Number of people clicked on a particular product or webpage.
  - The user location, usage of website duration, total number of sites visited and the user interface experience of the website.
  - What can this type of analytics tell us?
  - Aid in decision making and presenting the inferred information.
- Commercially, it is the collection and analysis of data from a website to determine that aspects of the website achieve the business objectives.

FIGURE 1.17   Web analytics.

1.6.1.9   *VISUAL ANALYTICS*

This type of analytics leverages the visualization techniques for presenting complex problems into simpler one (Figure 1.18).

1.7 BIG DATA ANALYTICS TOOLS WITH THEIR KEY FEATURES

Big data analytics program is commonly used to evaluate a broad data collection virtually. These software analytical methods help to recognize emerging developments in the industry, consumer expectations, and other data (Husamaldin and Saeed, 2020).

The cutting-edge big data analytics techniques have been the cornerstone to achieving effective data processing with the increase in big data volume and tremendous development in cloud computing. It will review the top big data analytics platforms and their main function.

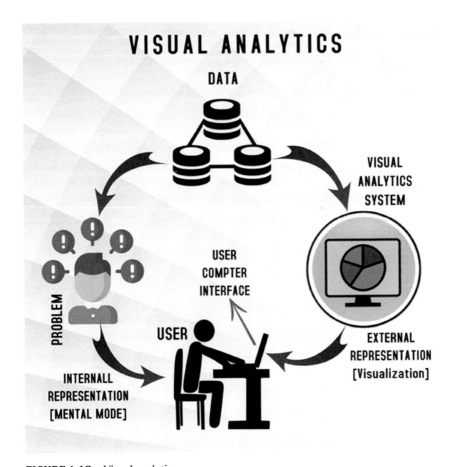

FIGURE 1.18 Visual analytics.

1.7.1   BIG DATA ANALYTICS TOOLS

1.7.1.1   APACHE STORM

Apache Storm is a large data computing system that is open sources and free. Apache Storm is also an Apache product that supports every programming language with a real-time application for data stream processing. It includes a distributed framework for real-time, fault-tolerant processing, with capabilities for real-time computation. About topology configuration, storm scheduler manages workload with multiple nodes and works well with the Hadoop Distributed File System (HDFS).

Features

- The transmission of 1 million 100-byte messages per second per node is calculated.
- Storm to guarantee that the data device is processed at least once.
- Great scalability in horizontal terms.
- Fault-tolerance built-in.
- Auto-reboot for crashes.
- Clojure-written writing.
- Fits with the topology of a direct acyclic graph.
- Output files are in the language of JavaScript Object Notation (JSON).
- Real-time analytics, log analysis, ETL, continuous computing, distributed RPC, deep learning have numerous use cases.

1.7.1.2   TALEND

Talend is a tool for big data that simplifies and automates the integration of big data. Native code is generated by its graphical wizard. It also supports convergence with big data, master data processing, and data consistency tests.

Features

Streamlines large data with ETL and ELT.

- Achieve the pace and size of the spark.
- Expedites the switch to real-time.

- Handling various sizes of info.
- Provides multiple connections under one roof, allowing it to tailor the solution according to its needs.
- By creating native code, Talend big data platform simplifies the usage of MapReduce and Spark.
- Smarter accuracy of data with machine learning and analysis of natural languages.
- To speed up big data projects, Agile DevOps.
- Streamline all the processes in DevOps.

1.7.1.3 APACHE COUCHDB

It is an open-sources, cross-platform, document-oriented NoSQL database that strives to render a modular design simple to use and manage. It is written in Erlang, a competition-oriented script. In JSON records, Couch DB stores data that can be downloaded electronically or queried using JavaScript. With fault-tolerant storage, it provides distributed scaling. Through specifying the Couch Replication Protocol, it facilitates accessing data.

Features
- CouchDB is a database with a single node that functions like every other database.
- It enables a single logical database server to be run on any number of computers.
- The ubiquitous HTTP protocol and the JSON data format are used.
- Inserting, updating, retrieving, and deleting documents is quite simple.
- It should convert the JSON file into different languages.

1.7.1.4 APACHE SPARK

Spark is also a very popular and open-source analytics platform for big data. Spark has over 80 operators at the highest level to render parallel applications simple to create. In a broad variety of organizations, it is used to process massive datasets.

Features

- It helps to operate an application in the Hadoop cluster, in memory up to 100 times faster and on disc 10 times faster.
- It provides fast-processing lighting.
- Sophisticated analytics help.
- Integration capacity with Hadoop and current Hadoop data.
- In Java, Scala, or Python, it provides optimized APIs.
- In-memory data processing capabilities are supported by spark, which is far quicker than the disc processing leveraged by MapReduce.
- Furthermore, spark operates on both cloud and on-prem HDFS, OpenStack, and Apache Cassandra, bringing another degree of flexibility to the business's big data operations.

1.7.1.5 SPLICE MACHINE

It is an analytics platform for large data. Their architecture, including AWS, Azure, and Google, is scalable across public clouds.

Features

- It can scale from a few to thousands of nodes dynamically to allow applications at every scale.
- Each query to the distributed HBase regions automatically evaluates the splice machine optimizer.
- Reduce management, deploy more rapidly, and decrease the risk.
- Create, evaluate, and deploy machine learning models by consuming fast streaming data.

1.7.1.6 PLOTLY

Plotly is a platform for visualization that helps users to build web-sharing maps and dashboards.

Features

- Convert some details easily into eye-catching and insightful graphics.

- It offers fine-grained knowledge regarding data provenance to audited industries.
- Via its free city plan, Plotly provides unregulated public file hosting.

1.7.1.7   AZURE HDINSIGHT

It is a cloud-based spark and Hadoop operation. It offers cloud offerings of big data in two types, regular and luxury. It provides the company with an enterprise-scale cluster for operating the big data workloads.

Features
- With an industry-leading SLA, efficient analytics.
- It provides protection and control for enterprise-grade.
- Safe computer properties and apply for access and compliance measures on-site to the cloud.
- For developers and researchers, a high-productivity platform.
- Integration of leading technologies for efficiency.
- Without buying new hardware or paying other up-front expenses, deploy Hadoop in the cloud.

1.7.1.8   R

R is a computer language and free software, and figures and images are computed here. The R language is popular for the development of statistical software and data analysis among statisticians and data miners. Many statistical tests are provided by the R Language.

Features
- R is often used to facilitate wide-scale statistical analysis and data visualization along with the JupyteR stack (Julia, Python, R). JupyteR is one of the fit commonly used big data visualization software, 9000 plus Comprehensive R Archive Network algorithms and modules that enable any analytical model running it to be composed in a convenient setting, modified on the go, and examined at once for the effects of the study. Language R is as follows:
- Within the SQL server, R will run.

- On all Windows and Linux servers, R runs.
- Apache Hadoop and Spark Helps R.
- R is incredibly compact.
- R scales effortlessly from a single test computer to large data lakes on Hadoop.
- Effective facility for data handling and retrieval.
- It offers a collection of operators, in specific matrices, for calculations on arrays.
- This offers a cohesive, organized set of data processing large data resizes.
- It offers graphical data processing facilities that view either on-screen or hardcopy.

1.7.1.9   SKYTREE

Skytree is a tool for big data analytics that empowers data scientists to build more precise models more quickly. It provides accurate models of predictive machine learning that are quick to use.

Features

- Algorithms strongly flexible.
- Computer scientists' artificial intelligence.
- It helps data scientists to grasp and interpret the reasoning behind ML decisions.
- Fast to use the interface or assists in programmatically using Java. SkyTree.
- Interpretability of model.
- It is intended to resolve robust predictive concerns with data preparation capabilities.
- GUI and programmatic access

1.7.1.10   LUMIFY

Lumify is considered a platform for visualization, big data fusion, and analytics. It allows users with a suite of analytical options to uncover links and explore relationships in their results.

Features

- It offers some automated formats for both 2D and 3D graph visualizations.
- Connects research between graph organizations, mapping systems integration, geospatial research, multimedia analysis, collaboration in real-time across a collection of projects or workspaces.
- Relevant intake processing and design components for textual information, photographs, and videos are included.
- This space mechanism helps it to arrange work into a variety of tasks or workspaces.
- It is focused on tested, scalable large data technologies that are focused on
- Supports the ecosystem centered on the cloud. It fits for Amazon's AWS well.

1.7.1.11 HADOOP

The long-standing big data analytics leader is recognized for its capabilities for large-scale data processing. Thanks to the open-source big data system, it has low hardware specifications and can operate on-premises or in the cloud. The key advantages and characteristics of Hadoop are as follows:

- Hadoop Distributed File System, targeted at HDFS (Huge-Scale Bandwidth).
- An extremely configurable large data analysis model—(MapReduce).
- A Hadoop resources control resources scheduler—(YARN).
- The required glue to enable modules from third parties to function with Hadoop (Hadoop Libraries).

It is intended to scale up from Apache Hadoop, a distributed file system, and big data handling program platform. Using the programming model MapReduce, it processes datasets of large data. Hadoop is an open-source application written in Java that offers support for cross-platforms.

This is, without a doubt, the top big data platform. Hadoop is employed by over half of the Fortune 50 firms. Some of the major names like single servers on thousands of computers for Amazon Web services, Hortonworks, IBM, Intel, Microsoft, Twitter, etc.

Features

- Enhancements to authentication when using HTTP proxy servers.
- Hadoop compliant file system effort definition.
- Help for extended attributes in the POSIX-style file system.
- It creates a strong atmosphere that is well-tailored to the theoretical needs of a developer.
- This adds versatility to the collection of data.
- It makes it easier to process data more rapidly.

1.7.1.12   QUBOLE

Qubole data service is an autonomous and all-inclusive big data platform that manages, learns, and optimizes its use on its own. This helps the data team to focus instead of maintaining the platform on company performance. Warner music community, Adobe, and Gannett are among the very, very few popular names that use Qubole. Revulytics is the nearest rival of Qubole.

1.8   TECHNIQUES OF BIG DATA ANALYSIS

A collection of numbers or characters is raw data or unprocessed data. For the extraction of decision-based material, data is sampled and analyzed, while expertise is extracted from a significant amount of experience in the subject (Hazen et al., 2014). Wherever an entity or atmosphere alters, sensory experience exists. Social knowledge, the creation of rules, practices, beliefs, rituals, social functions, icons, and languages, offers citizens the skills and behaviors they need to engage in their own cultures.

Phenomena in theory end up suggesting that what is perceived as the foundation of truth. Phenomena are also understood to be phenomena that appear or that human beings witness. A phenomenon can be an observable attribute and occurrence of matter, electricity, and the motion of particles in physics.

Data on actual phenomena may be viewed as perception or knowledge. Geographical evidence is thus known as an observation of physical and human spatial phenomena. Geographical big data processing is targeted at investigating the complexities of geographical reality (Russom, 2011).

The characteristics of large data analysis are derived from the characteristics of large data in the form of data structural collection and structural analysis. Therefore, in Figure 1.19, six big data analytics approaches are suggested.

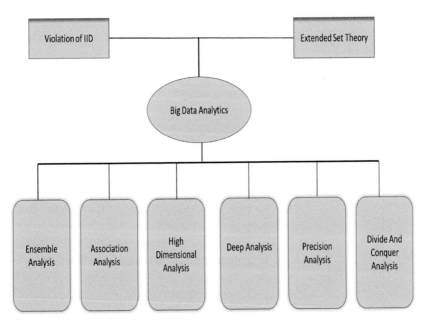

FIGURE 1.19   Two theoretical breakthroughs and six techniques in big data analytics.

1.8.1   ENSEMBLE TECHNIQUE

Ensemble data processing, also known as multidataset processing or multialgorithm analysis, is performed on the whole data array or a large volume of data. Big data is defined as the entire data collection without any sampling purpose. What does the entire kit entail?

Data from resampling, classified and unlabeled data, previous data, and subsequent data can all be used loosely. The term "ensemble" is understood in the context of ensemble learning in machine learning, the ensemble approach in statistical mechanics, and the Kalman ensemble filtering in data assimilation (Dey et al., 2018).

The whole data is sometimes split into three sections, that is, training data, test data, and validation data, in supervised learning and image recognition. The picture classifier is learned and applied to the test data using training data. Lastly, the findings of picture classification are checked by ground reality. Generally, on training data rather than test data, the classifier produces reasonable output.

To accomplish a tradeoff between the complexity of the model and the classification outcome on training data, ensemble analysis is used to produce the global optimum result on the whole set of training data, test data, and validation data. Structural risk minimization is used in support vector machine learning to address the problem of overfitting, that is, the learning model is closely adapted to the particularities of training data but badly applicable to test data.

1.8.2   ASSOCIATION TECHNIQUE

Big data is typically obtained without special methods for sampling. Since data developers and data users are typically somewhat different, the cause-and-effect relationships hidden in scientific data are not clear to individual data consumers. The set theory, that is, the theory about participants and their relationships in a set, is general enough to deal with data analysis and problem-solving.

In certain cases, the big data association (Wiech et al., 2020) is linked to the relationship of set members. Association research is critical for multisiting, multitype, and multidomain data processing. Data mining with association rule algorithms, data association in goal monitoring, and network relation research are all examples of connection research.

There are two elements of association law, an antecedent (if) and a consequent (if) in the Apriori algorithm of association rules mining (then). Through analyzing data and using the support and trust criteria to identify the value of relationships, association guidelines are developed to define certain if/then patterns. Help is a metric that indicates how many items are available in the purchasing catalog.

Trust is a measure of how much it has been found that the if/then claims are valid. Association rules mining is an example of transaction data calculation of conditional likelihood P (consequent/antecedent). However, unlike the asymptotic study of mathematical estimators, the usefulness of the rules relating to mined associations depends only on the thresholds of assistance and trust that users manually provide (Stucke and Grunes, 2016).

Typically, numerous sensors are used in object detection and multiple objects are traveling in a dynamic setting. Three associations of calculation (observation) data association, goal state association, and goal-to-measurement association need to be studied to define and monitor goals.

The most probable future goal position calculation is used to adjust the goal's state estimator to enable the tracking device to work properly. A distance function from the expected state to the observed state is the likelihood of the given calculation is right.

Kalman filtering is a time-varying recurrent estimator or filtering of a time series that is ideal for processing data streams in real-time. Kalman filtering, also known as linear quadratic estimation or trajectory optimization, is a time-based algorithm that produces estimates of unknown states (variables) from a set of noise-disturbed measurements (observations) that tend to be more reliable than single-calculation methods.

The covariance of state error and operator of state prediction is regarded as stochastically and deterministically as state associations. Similarly, the covariance and observation operator of measurement loss are considered stochastically and deterministically as measurement relations (Mithas et al., 2013). In a way, the association of data and the pace of big data is defined by the Kalman filtering.

Social people and social interactions form a social network of social networking. Concepts and logical connections constitute a cognitive network of natural language or mental models. The associations of networks, often called connected data, are typically formalized by social ties and logical relations. The network, in essence, is a topological construct.

The network relation defines the community or community, which may also be considered an alliance. The interaction in big data is essentially known to be the partnership in set theory. Associations are formalized as advanced partnerships as metrics are allocated. Statistical connections and geometric relation, for example, are correlations eventually assigned to the probability metric and geometric metric.

1.8.3 HIGH-DIMENSIONAL ANALYSIS

Big data implies a large variety of data. The mathematical space (object) dimension is defined as the minimum number of coordinates required to specify any point within it, and the vector space dimension is defined as the number of vectors on any space basis or the number of coordinates required to specify any vector in mathematics.

The material dimension is a property of the material that is independent of the space in which it is embedded. The dimension refers to the number of points of view from which the physical world can be discerned.

Classical mechanics explains the physical world in three dimensions. Starting from a certain point in space, the basic ways in which to pass are up/down, left/right, and forward/backward. Concerning time, the equations of classical mechanics are symmetric (Elahi, 2019). Time is directional in thermodynamics, concerning the second thermodynamic theorem, which specifies that the entropy of an independent system never decreases.

Such a structure progresses naturally into a thermodynamic equilibrium of maximum entropy states. Multidimensional analysis for a sample (realization) of multiple random variables is available in statistics and econometrics. In classical multivariate review, high-dimensional statistics analyze data whose scale is greater than certain measurements. Even the data vector component can be greater than the sample size in certain implementations.

Usually, in big data analysis, the curse of dimensionality emerges. Once the dimensionality increases, the space volume increases so rapidly that the data available become sparse. This sparsity will lead to the statistical error of significance and the dissimilarity between objects in high-dimensional space when handling traditional data algorithms.

This paradox does not exist in low-dimensional space; however, it only occurs in high-dimensional space, which is known as the dimensionality curse. Using distance or variance metrics, dimension reduction is a way of minimizing the number of random variables by choosing a subset of the original variables (called subspace) to conserve the volatility of the original variables as much as possible. There exist linear and nonlinear transforms of dimension reduction. The primary component analysis, for example, is a linear transformation of dimension reduction.

1.8.4 DEEP ANALYTICS

Today, the amount of evolving data is high enough for difficult training on artificial neural networks. In the meantime, multicore CPU, GPU, and FPGA high-performance computing systems dramatically decrease the training period of deep artificial neural networks. Typical artificial neural networks are dynamically enhanced in such cases, with hidden layers of latent variables, and so-called deep learning, as opposed to shallow learning, is created.

It is often assumed that as its understanding grows rich, human awareness of the natural world is rising deeper. The opportunity for deep research is to discover the dynamic structural properties of big data. Via deep research, unobservable variables, secret factors, secret hierarchies, contextual associations, and the dynamic distribution of random variables can be identified. The latent Dirichlet allocation models, the deep Boltzmann machine, the deep belief net, the hidden Markov model, and the Gaussian mixture model typically exist.

Multilayered artificial neural networks are used to create deep learning architectures. In a general artificial neural network, there are numerous nonlinear neuron processing units (activation units), such as the sigmoid function. Different neurons from each layer are viewed as multiple variables in principle. Deep analysis is considered as a deterministic (or stochastic) translation function from input to output.

Theoretical mechanisms for developing multilayered artificial neural networks can be used as theoretical approximation models, data-fitting functions, and maximum likelihood estimators. In complex data modeling, the multilayered artificial neural network, which has multiple layers, multiple variables, and nonlinear activation units, is very successful.

However, for deep learning of some dynamic functions, there are no universal theories of approximation available so far. The output of different architectures, which are not always tested on the same datasets, cannot be compared all together.

In multivariate multilayered neural networks that are completely connected or sparsely linked, multiple parameters must be determined, including the number of layers, the number of units per layer, the weights, and thresholds of each activation unit.

By finding a maximal answer to the data appropriate function or mathematical estimator, this will increase the issue of ill-posed problems and low measurement performance. Overfitting and expensive computing time are typical outcomes of the complicated paradigm. To solve these fitting issues, regularization techniques are developed.

To boost the performance, several numerical algorithms have been developed, such as pretraining, concave-to-convex approximation, and computing the gradient on several training datasets at once.

1.8.5   PRECISION METHODS

Precision is the resolution of the representation, which is often determined by the number of decimal or binary digits. Accuracy is the similarity of a calculation to the true value of numerical analysis. Instead of detail and accuracy, bias and heterogeneity are included in the statistics.

Bias is the amount of inaccuracy, and heterogeneity is the quantity of imprecision. In fact, the truth of a series of calculation outcomes is the average's closeness to the actual (true) value, and accuracy is the consistency of a set of results. Ideally, with measures all like and closely grouped around the true meaning, a measuring system is both exact and reliable.

A measuring method, one or both, maybe exact but not exact, is reliable but not precise. Increasing the sample size usually improves precision when an experiment involves a systemic mistake but does not boost precision. The removal of systemic errors increases precision but does not modify accuracy. A precision review is used from the viewpoint of data utility and data accuracy to determine the veracity of data.

1.8.6   DIVIDE-AND-CONQUER TECHNIQUE

Divide-and-conquer is a mathematical methodology for improving problem-solving efficiency and the speed at which large data is analyzed. Divide-and-conquer analysis is used to recursively separate a problem into two or three subproblems in the stage of division, until they become simple enough to be solved automatically in the stage of conquest. When both subproblems have been solved, they are combined into a single answer to the main problem.

Distributed computing, such as cloud computing and distributed information computing, can be thought of as divide-and-conquer in space, whereas parallel computing (multicore computing and cluster computing) can be thought of as divide-and-conquer in time.

In terms of unstructured data and time limits, steam generation and real-time computing are a divide-and-conquer calculation. From the perspectives of virtualized networks and culture of mind, cloud computing and interactive information computing are more akin to divide-and-conquer computing.

In a shared-memory multiprocessor, high-performance concurrent processing utilizes parallel algorithms, whereas distributed algorithms are used to coordinate a large-scale distributed machine. In multiprocessor computers, divide-and-conquer analysis is quickly executed, and separate subproblems may be executed on different processors.

By transmitting messages between the processors, information is shared. To solve a query, concurrent computation requires several computational elements concurrently. This is done by splitting the issue into different components such that each processing component will perform the algorithm component concurrently with the others.

The processing components may be complicated, requiring the resizing of multiple processors and many networked computers, such as a single computer. After frequency scaling hit a ceiling, multicore systems became more prevalent and concurrent algorithms became widespread.

A multicore processor is a single-chip processor with multiple processing units (cores). The main memory of a parallel computer is either shared memory (all processing elements share a common address space) or distributed memory (each processing element has its own local address space).

Parallel processing is a method of computing in which many operations are carried out at the same time, allowing complicated problems to be broken down into smaller bits and solved at the same time. Parallel computing is made up of many layers of bit, instruction, input, and task parallelism.

These subtasks will then be performed by the processors concurrently and mostly cooperatively. OpenMP is one of the most frequently used shared memory application programming interfaces (APIs) in programming languages, although the message passing interface is the most frequently used API for message-passing applications. A concurrent computation programming framework is MapReduce in the machine cluster. Concerning data organized in (key, value) pairs, the map and reduce functions of MapReduce are both specified.

All maps may be done in tandem, providing that each mapping process is independent of the others. At the same time, all outputs of the map operation which share the same key are presented to the same reducer. Then, the reduce function is added to and category in parallel.

However, MapReduce differs from relational databases in that it lacks a structural data schema and allows for quick evaluation of B-tree and hash partitioning techniques, and it is mostly used for simple or one-time retrieval tasks (Elahi, 2019).

The data processing date, time constraints, and time validity are all important in real-time computation. A real-time database is a database system that manages constantly changing workloads using real-time computation. This differs from traditional databases of permanent records in that it remains mostly intact over time. A stream is a series of data or a sequence of operations. Nonstructural data and infinite-size continuous data are other terms for streaming data.

Unlike archived data, stream data is usually a real-time, read-only sequence. For the streaming data portion, incremental update, and Kalman filtering, the time window of data processing has implications for the data sampling rate and underlying temporal similarity of windowed data.

Both nodes are on the same local network in cluster computing and use identical hardware. A computer cluster is a community of computers strongly coupled that operate tightly together such that they can be viewed as a single computer in certain respects. Multiple standalone computers linked by a network are comprised of clusters.

Nodes are shared around globally and administratively dispersed networks of grid computing, and they utilize more heterogeneous hardware. The most distributed method of parallel computing is grid computing, and it allows the use of machines that connect across the Internet to operate on a given topic.

In distributed computing, a problem is divided into several operations, each of which is resolved by one or more machines that communicate by exchanging messages, and each computer has its own private memory (distributed memory).

Computational structures are referred to as devices or nodes. A distributed structure of databases consists of locations that are closely connected and do not share physical components. The distribution is translucent, that is, as though it were one abstract structure, consumers can communicate with the system.

Also, transactions are open, ensuring that each transaction must preserve the credibility of the database across several databases. Cloud computing, also known as on-demand computing, is a form of Internet-based computing that delivers pooled processing services and data to computers and other devices on-demand, allowing for ubiquitous, on-demand access to a common pool of configurable computing resources.

Compared to conventional parallel computing methods, cloud computing offers the resizes and technology to create data/computation-intensive parallel applications at far more accessible rates. Virtualization is the primary supporting technology for cloud computing. Virtualization software splits a real computer machine into one or more "virtual" units, each of which can be used and managed to perform computing tasks.

Service-oriented architecture concepts are used in cloud computing to help consumers break down problems into services that can be integrated to provide a solution. According to numerous models, cloud-computing companies provide their "packages," which shape a stack, that is, infrastructure, platform, and software as a service.

The human–machine–environment interface mechanism can be loosely conceptualized in distributed intelligence computing to be a mechanism of agents. Individuals' rationality is limited by the intelligence available, the tractability of the judgment question, the cognitive limitations of their brains, and the time available to make the decision, according to Simon's theory of bounded rationality.

A rational agent is one who has simple expectations, models ambiguity using variables' expected values or functions, and frequently chooses the action with the best-expected effect for himself from among all possible actions. Something that makes choices, usually a human, computer, or software, maybe a logical agent.

An intelligent agent is an autonomous entity that observes and acts on an environment utilizing actuators by sensors and guides its behavior toward achieving objectives. Intelligent agents can often learn to accomplish their objectives or use intelligence.

1.8.7 GEOGRAPHICAL BIG DATA METHODS

Six big data analytics methods are still applicable for spatial big data. The unique thing about spatial big data is the spatiotemporal association, such as geometric associations, mathematical similarities, and semantic connections, compared to general big data.

Geometrical calculation research, human spatial data analysis, and physical spatial analysis may be loosely grouped into spatiotemporal data analysis. The artificial planet, smart planet, and earth simulator analytics are partially suggested by three kinds of study.

The human–machine–environment scheme is gradually being encountered in digital form through the creation and implementation of device and communication technology. A vast amount of spatial data, such as satellite imagery of earth observation and data from the mobile Internet, is gathered jointly or independently.

In certain instances, independent of technology domains and sampling methods, spatial big data is virtually the whole data collection. Ensemble data processing is commonly applied to large data in geography.

Geometrical relations, mathematical associations, and semantic relations are classified into spatiotemporal relations. Space and time in mathematics may be formalized into geometric quantities. Measurement modification in surveying, for example, is the least square measurement of angles, corners, and elevations of geometrical features (parameters).

The duration is scaled by the velocity of objects "motion, such as heavenly bodies" timing objects and atom clocks. Currently, for a vast number of georeferenced details, physical and human geographical phenomena are detected utilizing different survey tools.

In the modern environment, the machine digitally tracks topographical observation and geographical phenomenon sensing, and geometrical interaction analysis in geographical big data analytics is stressed.

KEYWORDS

- **big data**
- **structured data**
- **unstructured data**
- **volume**
- **velocity**
- **variety**
- **veracity**

REFERENCES

Andrejevic, M.; Gates, K. Big Data Surveillance: Introduction. *Surveillance Soc.* **2014,** *12* (2), 185–196.

Anuradha, J. A Brief Introduction on Big Data 5Vs Characteristics and Hadoop Technology. *Procedia Comput. Sci.* **2015,** *48,* 319–324.

Dey, N.; Hassanien, A. E.; Bhatt, C.; Ashit, A.;Satapathy, S. C., Eds. *Internet of Things and Big Data Analytics toward Next-Generation Intelligence*; Springer: Berlin, 2018; pp 3–549.

Elahi, I. *Scala Programming for Big Data Analytics: Get Started with Big Data Analytics Using Apache Spark*; Apress, 2019. ISBN: 9781484248102.

Engelbrecht, E. R.; Preez, J. A. D. Introduction to Quasi-open Set Semi-supervised Learning for Big Data Analytics, 2020.

Goul, M.; Sidorova, A.; Saltz, J. Introduction to the Minitrack on Artificial Intelligence and Big Data Analytics Management, Governance, and Compliance, 2020.

Hazen, B. T.; Boone, C. A.; Ezell, J. D.; Jones-Farmer, L. A. Data Quality for Data Science, Predictive Analytics, and Big Data in Supply Chain Management: An Introduction to the Problem and Suggestions for Research and Applications. *Int. J. Product. Econ.* **2014,** *154,* 72–80.

https://searchbusinessanalytics.techtarget.com/definition/big-data-analytics

https://www.sas.com/en_in/insights/analytics/big-data-analytics.html

Husamaldin, L.; Saeed, N. Big Data Analytics Correlation Taxonomy. *Information* **2020,** *11* (1).

Kankanhalli, A.; Hahn, J.; Tan, S.; Gao, G. Big Data and Analytics in Healthcare: Introduction to the Special Section. *Inform. Syst. Front.* **2016,** *18* (2), 233–235.

Kim, H.; Pries, R. D. *Big Data Analytics*; Auerbach Publications, 2015. ISBN: 9781482234527.

Loshin, D. *Big Data Analytics*; Morgan Kaufmann, 2013. ISBN: 9780124186644.

Michael, K.; Miller, K. W. Big Data: New Opportunities and New Challenges [Guest Editors' Introduction]. *Computer* **2013,** *46* (6), 22–24.

Mithas, S.; Lee, M. R.; Earley, S.; Murugesan, S.; Djavanshir, R. Leveraging Big Data and Business Analytics [Guest Editors' Introduction]. *IT Professional* **2013,** *15* (6), 18–20.

Renuka Devi, D.; Sasikala, S. Feature Selection and Classification of Big Data Using MapReduce Framework. In *Intelligent Computing, Information and Control Systems. ICICCS 2019. Advances in Intelligent Systems and Computing*; Pandian, A., Ntalianis, K., Palanisamy, R., Eds., Vol. 1039; Springer, Cham, 2020.

Russom, P. Big Data Analytics. *TDWI Best Practices Report*, Fifth Quarter **2011,** *19* (4), 1–34.

Saltz, J.; Sidorova, A. Goul, M. Introduction to the Minitrack on Artificial Intelligence and Big Data Analytics Management, Governance, and Compliance. In *Proceedings of the 53rd Hawaii International Conference on System Sciences*, 2020.

Stucke, M. E.; Grunes, A. P. *Introduction: Big Data and Competition Policy. Big Data and Competition Policy*; Oxford University Press, 2016.

Tsai, C. W.; Lai, C. F.; Chao, H. C.; Vasilakos, A. V. Big Data Analytics: A Survey. *J. Big Data* **2015,** *2* (1), 1–32.

Wiech, M.; Boffelli, A.; Elbe, C.; Carminati, P.; Friedli, T.; Kalchschmidt, M. Implementation of Big Data Analytics and Manufacturing Execution Systems: An Empirical Analysis in German-Speaking Countries. Production Planning & Control, 2020; pp 1–16.

CHAPTER 2

Preprocessing Methods

ABSTRACT

This chapter outlines the need for preprocessing data and various methods in handling the same. Both text and image preprocessing methods are also highlighted. As the data are collected from various resources, it might contain discrepancies and inconsistencies in data representation. To overcome the same, the pre-processing methods used are data cleaning, data integration, data transformation, data reduction and data transformation. The various pre-processing algorithms under each category are discussed with a neat procedure and examples. The challenges of big data stream processing are also highlighted. At the end of the chapter text and image pre-processing methods are discussed with examples for clear understanding.

2.1 DATA MINING—NEED OF PREPROCESSING

Before we discuss the challenges in data mining, let us understand the life cycle of Machine Learning (ML) process. The life cycle of any ML starts with data gathering or data acquisition. This is presented in Figure 2.1.

"Data" is vital for any data analytics process. Understanding data and preparing the data for further analytics is indeed considered to be a crucial step in mining. A core collection of (García et al., 2016) approaches are called as "data preprocessing," which is applied before the analytics process. This is the part of the knowledge discovery (Zaki and Meira, 2014).

Research Practitioner's Handbook on Big Data Analytics. S. Sasikala, PhD, D. Renuka Devi, & Raghvendra Kumar, PhD (Editor)
© 2023 Apple Academic Press, Inc. Co-published with CRC Press (Taylor & Francis)

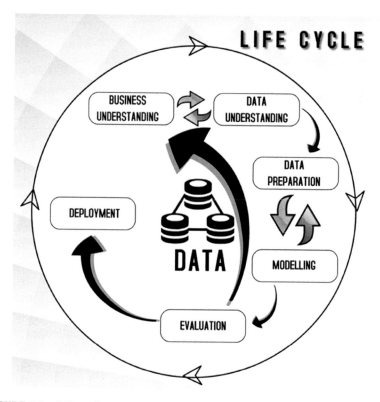

FIGURE 2.1   Life cycle.

As the data are collected from various resources, it might contain discrepancies and inconsistencies in data representation. Streaming data are collected at a very fast rate across verticals. When this kind of scenario emerges, it is obvious that an enhanced mechanism is required to handle them effectively. Thus, preprocessing algorithms have been proposed by various researchers. The objective of any algorithm is to:

- Handle—real-time data
  - Real-time data consists of noise
  - Incomplete information
  - Errors
  - Outliers
- Handle—streaming characteristics of data
- Handle—scalability issues

Issues and Challenges

Most of the common problems are listed below:

- Incomplete data
- Error due to manual typing
- Inconsistent data
- Raw data format
- Mismatch in representing numerical values
- Data-type error
- Crashed file
- Irrelevant data

Incomplete Data

Not all data are complete; some datasets will have junk values or incomplete information. Different approaches are proposed for handling those types of issues. First, we have to explore the causes of the missing data and find the reasons behind that.

This is usually identified from the following perspectives:

- The random nature of missing is that the missed data has no relation with any of the data points present in the dataset. There is no dependency of missed value with the rest of the data present. The entry is simply missed out or not captured that is why it is not recorded.
- Second perspective is that, still the random nature of value is missing but it has some relevance with the existing data. Some point of relevancy exists between them. Some values are purposely missed out based on the other data values.
- In the third one, there is a complete dependency that exists and apparently the data is omitted.

The missed values can be treated for all aforesaid cases. If no relevance exists, we can very well omit the data, but if there is relevance then deletion and omission can lead to anomalies. When we try to delete records that may less represent a particular class in the entire dataset. Another possible method is to replace the missed value with some other value based on statistical measure.

Error Due to Manual Typing

Data inconsistency arises when the user commits a typo error. For example, entering string values in the numeric data field or appending some special characters to the numeric value. To change this, either remove the special character or transform the non-numeric into numeric field.

Inconsistent Data

The variations in the data are called as "inconsistency." This is considered to be a major issue in preprocessing. For example, the same value is represented in two different forms but the meaning of both the terms is same.

For example, zero or 0 is the same, but different representations for the same field lead to inconsistency. For the entire datasets, the unique way of representation has to be followed for the same field. Providing more than one value to a single column also leads to inconsistency.

Raw Data Format

Data in all forms can be collected for analytics. When we collect the data in raw format, make sure that in which format it is to be represented. For example, Date of birth is represented in many formats, therefore decide on which format (dd/mm/yyyy or mm/dd/yyyy) that the date has to be represented in the dataset. Always check and verify the same for inconsistencies.

Mismatch in Representing Numerical Values

Numerical data can be expressed in various metrics like grams or kilograms, temperature in Celsius or Fahrenheit. The common representation of units is decided and it is uniformly represented across the dataset.

Data-type Error

Human error or incorrect representation of data can lead to inconsistency. The data representation is verified and checked before the analytics process.

Crashed File

When we try to import some files into native formats, the file may get crashed because of software we use for conversion. File manipulation is the biggest issue when we try to convert one file format into another.

Irrelevant Data

Some of the gathered data may contain sensitive information (like password, pin number, etc.), those data should be eliminated before the process.

2.2   PREPROCESSING METHODS

The various types of preprocessing methods are given in Figure 2.2.

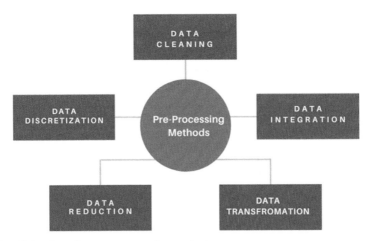

FIGURE 2.2   Overview—preprocessing methods.

2.2.1   DATA CLEANING

Cleaning of data is done by replacing the missed data with some values, removing noisy elements, detecting outliers, and inconsistencies. The objective of this process is to enhance the quality of data. The quality of data degrades when they are incorrectly represented. The noisy data reduces the accuracy of the ML model that leads to poor performance. Various reasons for unclean data are listed below:

- Data missing—absence
- Violation of rule or representation error
- Data integrity
- Data mismatch or represented in various format for the same column
- Wrong usage of primary keys

2.2.2.1   MISSING VALUE

Once the missing values are identified, it should be handled. Several techniques are available to handle the missing data, but the appropriate method is chosen based on the datasets. The other factors like specific domain, data mining process is also taken for consideration. The ways of handling missed data are:

- Ignore—Missing data is completely ignored. That is the row with the missing value is deleted. Deletion should not result in poor performance. This is very tedious and a time-consuming process. This method is not considered for large- scale problems. The deleted row should not completely ignore the significant features.
- Constant—Replace the missed value with a constant. For example, zero or NULL can be used to replace the missed values all over the dataset. Global hard encoding, one-to-one mapping techniques are existing for this purpose.
- Mean—Mean of the column values is used for replacing the null values. For example, if the salary column has the missed value, the mean of all salaries can be calculated and replaced uniformly to all the null values. Statistical measures can also be employed.
- Forward/backward fill—The preceding or succeeding values of the specific columns are used.
- Algorithm—Use specific an algorithm for predicting the missed value.

Algorithms

The overview of the algorithm used for handling "missing value" is presented below:

- **Statistical Method**
  - Ignoring method—List-wise deletion, Pairwise deletion
  - Missing data model—Maximum Likelihood, Expectation Maximum Algorithm

- **Machine Learning Method**
  - Imputation Techniques
    - Mean and Median
    - Regression
    - Multiple imputation
    - k-Nearest Neighbor
    - k-means
    - Fuzzy k-Mean

2.2.2.2 NOISY DATA

Noisy data can arise during typo error, data collection from sensors/ devices, and so on. For example, erroneous encoding for female is "E" instead of "F," negative value for the numeric field where it has to be positive (Salary = −20000) and so on.

Noisy data are well handled by the binning techniques. The given data is sorted and separated into number of bins. Partitions of the bins are done with equal width and height. Once they are segregated, smoothening is applied. In this way, outliers are identified and removed. Box-plot method is also very helpful in identifying the outliers.

The various types of noises are depicted in Figure 2.3.

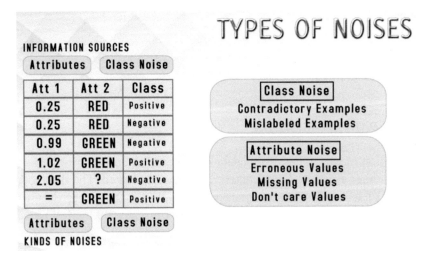

FIGURE 2.3 Types of noises.

From Figure 2.3, we can infer the following:

Class noise occurs when the output class label is incorrect (Mislabeled) and attribute noise occurs when there is an attribute error/missing value error.

The various filters are used to remove the noises,

- Ensemble-based filtering
- Iterative-based filtering
- Metric-based filtering

The list of other approaches is presented in Figure 2.4.

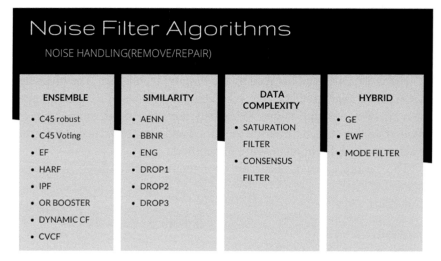

FIGURE 2.4 Noise filter algorithms.

2.2.2 DATA INTEGRATION

Integration means collations of data from various sources. The format of data is either structured, unstructured, or semistructured. The data source can be of a single type or heterogeneous. When we gather data from multiple sources the following issues may arise. They are the following:

- Merging different schema formats into combined format is a challenging task. For example, the Employee name is "empname" in database, the same column is represented by "Ename" in another schema definition. When we integrate, we have to resolve this issue and establish a single representation for the employee's name column.
- Data representation in different formats is verified and uniform representation is devised. For example, weights are represented by grams or kilograms. The conflict has to be resolved and appropriately represented.
- When we integrate, there is a possibility that the same feature is repeated, leading to redundancy issues.

2.2.3 DATA TRANSFORMATION

This is conversion of data into other forms, like aggregation or generalization of datasets. They are the following:

Normalization

Normalization is a scaling technique that transforms the data into rage between 0 and 1. But the originality of the data remains the same even though after transformation. This method is applied when the attributes are on different scales. The widely used methods are as follows:

- Z normalization
- Min–max normalization
- Mean normalization

Z Normalization

Z normalization is also called "standardization." This can done by the following equation:

$$Z = \frac{x - \mu}{\sigma}$$

Z = standard score
x = observed value
μ = mean of the sample
σ = standard deviation of the sample

Min–Max

Min–max scales the given value into a range [0,1]. The equation is given as

$$x' = \frac{x - min(x)}{max(x) - min(x)}$$

Min and max are the minimum and maximum value in the dataset, respectively.

Mean

The mean of the values is used in the equation. And the transformations is done by the equation

$$x' = \frac{x - average(x)}{max(x) - min(x)}$$

2.2.4   DATA REDUCTION

Data reduction is done when the dataset is large, and not all the features are significant for data analysis. Thus, reduction allows selecting only the optimal features by reducing the total number of features significantly. This way reduction enhances the mining process with only imperative features. Let us discuss the techniques of reduction as follows:

- Data cube aggregation
- Dimensionality reduction
- Data compression

2.2.4.1   DATA CUBE AGGREGATION

Aggregation means combining the data in a simpler form. For example, if the dataset contains sales data for a year, but only a quarterly sales report is to be presented then we can collate the three months data into one. We can summarize and present the data.

2.2.4.2   DIMENSIONALITY REDUCTION

This is the process of removing features, thereby reducing the dimensions of the dataset. The methods used for the selection process are forward selection, backward selection, and so on. When the features are reduced, make sure that it should not eliminate important ones.

Forward Selection

Forward selection starts with an empty set, features are added incrementally to form the final subset (Figure 2.5).

Backward Selection

In this method, all the given attributes are taken into consideration. At every iteration, the features are removed one by one thus producing the condensed subsets (Figure 2.6).

FORWARD SELECTION

INSTANCES :: {S1, S2 ,S3 ,S4 ,S5 ,S6 }

ATTRIBUTE FORWARD SELECTION

STEP 1: {} EMPTY SET

STEP 2: {S1}

STEP 3: {S1,S2}

STEP 4: {S1,S2,S6} - THE FINAL ATTRIBUTE SET

FIGURE 2.5   Forward selection.

BACKWARD SELECTION

INSTANCES :: {S1, S2 ,S3 ,S4 ,S5 ,S6 }

ATTRIBUTE BACKWARD SELECTION

STEP 1: {S1, S2 ,S3 ,S4 ,S5 ,S6 }

STEP 2: {S1, S2, S3,S4 ,S5}

STEP 3: {S1,S2 ,S3}

STEP 4: {S1, S2} - THE FINAL ATTRIBUTE SET

FIGURE 2.6   Backward selection.

Feature Selection

Feature selection (FS) is to select the significant features thereby reducing the extraneous and superfluous features with the objective to enhance the performance of machine learning model (Hall and Smith, 1998). The derived subsets have to maintain the originality and properly describe the characteristics of the given dataset.

The FS process may introduce generalizations and overfitting issues. Generally, in big data analytics, FS plays a vital role because reducing the feature space is an important and challenging task.

2.2.4.3 DATA COMPRESSION

Compression is the process of reducing the size of the features for the sake of the mining process. It can be either lossy or lossless. In the lossy approach, the quality is compromised during the compression process, wherein lossless preserves the quality of data after compression.

Lossless Compression

Encoding approaches are commonly used for lossless compression. This algorithm maintains the precise form of the original data even though they are converted into another form.

Lossy Compression

Transformation techniques such as wavelet transformation, principal component analysis are employed for compression. The compressed data may vary from the original but useful for analysis for large-scale problems.

2.2.5 DATA DISCRETIZATION

For any data analysis, domain and type of data should be known before applying any data-mining algorithms. The types of data such as categorical, nominal, and discrete must be handled in a specific way. For example, the categorial value is considered important in node split-up in a decision tree.

If the data is continuous in nature, then discretization of data is done priorly before the induction process begins. This is done by splitting the attributes in a range of continuous values and giving them separate labels for the same. This way it is used to make the mining process easier.

From the example given in Figure 2.7, the age attribute values are split into three groups and each group is given labels to identify. The discretization types are top-down and bottom-up. The techniques for discretization are given below:

- Histogram—It is a plot that represents the frequency distribution of data in continuous form. The distribution of data is depicted in such a way that we can understand the distribution pattern very well.
- Binning—The continuous data is divided into number of bins
- Cluster analysis—Grouping the similar data together
- Decision tree analysis—Building decision tree based on the node split-up criterion
- Partitioning datasets—Equal width and depth partitioning

Data Discretization

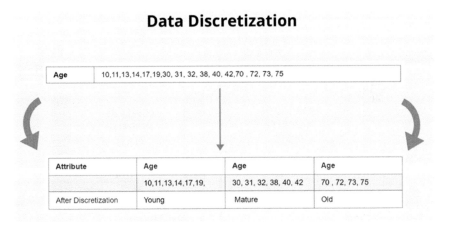

FIGURE 2.7 Discretization process.

2.2.6 *IMBALANCED DATA—OVERSAMPLING AND UNDERSAMPLING*

The class imbalance (López et al., 2013) occurs in data mining, where there is less representation of records for a particular class. If it is not treated may give rise to overfitting or underfitting difficulties. The analysis results of the imbalanced datasets are quite often biased toward the negative class, and then ultimately model prediction may result in higher misclassification error rate.

The datasets that contain more samples are called majority class and less representation of any class is called as a minority. The cost of misprediction is complex in minority classes compared with majority classes. Multiclass imbalance can happen when a dataset represents all the instances that belong to one of many target classes. It is the combination of several minority and majority classes.

To overcome this issue, resampling methods are employed for suitable data balance. Resampling is achieved by shuffling and rearranging the given datasets to maintain the balance. Sampling techniques are categorized as undersampling and oversampling (Tanha et al., 2020).

The two methods available (Figure 2.8) for class imbalance issues are:

- Undersampling—Subset of features are created by removing the majority features
- Oversampling—Features are created by replicating the existing (minority) features

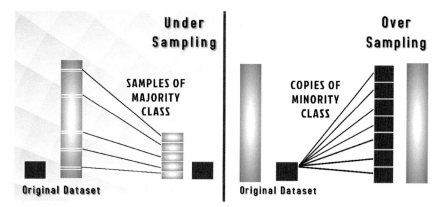

FIGURE 2.8   Undersampling and oversampling.

Undersampling is applied to reduce the number of occurrences of majority classes. The advantage of this method is less computational complexity and less memory requirements. Random sampling is one of the commonly used methods for handling undersampling. The disadvantage of undersampling is that sometimes removal of data instances might remove the significant records as well.

Oversampling is done by replicating some data instances in such a way that class balance is maintained. The main advantage of this method is that we retain all the records at the same time; repetition might lead to computational complexity. The simplest method of oversampling is random oversampling which randomly chooses the minority class instances and duplicates them. The disadvantage of this method is overfitting.

Oversampling is applied to minority classes by SMOTE (Synthetic Minority Oversampling Technique). SMOTE technique handles the imbalance between majority and minority classes and also removes noise (Pan et al., 2020).

When SMOTE (Figure 2.9) is applied, the synthetic samples are generated for minority classes. The working procedures are as follows:

Procedure
- A random data from the minority class is chosen
- The k-nearest neighbors from the data point are created
- Synthetic data is generated between a random point and one of the nearest neighbor points
- This procedure is repeated for the number of times, until we have an equal proposition of minority and majority classes

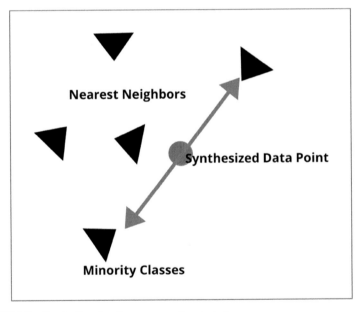

FIGURE 2.9 Synthetic minority oversampling technique.

2.3 CHALLENGES OF BIG DATA STREAMS IN PREPROCESSING

The fundamental characteristics of Big Data are already presented in Chapter 1. To extend the same, the challenges of preprocessing methods for big data streams are discussed in this section. Various preprocessing approaches for handling big data have been proposed. The challenges are outlined below,

- Conventional approaches are mostly full batch loading of features, but streaming features are incremental; they have to be processed immediately once they are gathered. Automated approaches are indeed required to process the features as soon as they arrive and fine-tune the parameters as well.
- Importantly, all the updation has to be synchronized with the model. Otherwise, the predictive model cannot accommodate the changes made upon the arrival of new features.
- To apply this change to the machine learning model, increment learning approaches have been proposed for updating the induction model incrementally (Renuka Devi and Sasikala, 2019).

The most of the contributions for preprocessing methods in big data analytics is based on MapReduce, Apache Spark, Flink (García et al., 2016) Twister, (Massively parallel feature selection) MPI, and MATLAB. The feature selection methods are elaborately discussed in Chapter 3.

The various preprocessing methods for big data streams processing are presented in Figure 2.10. Most of the approaches are based on parallel MapReduce-based system distributed across the nodes.

2.4   PREPROCESSING METHODS

2.4.1   TEXT PREPROCESSING METHOD

The main components of text preprocessing are:

- Tokenization
- Normalization
- Substitution
- Noise—removal

Tokenization

Tokenization is a process of dividing the given texts into number of smaller parts called "tokens." The large sentence taken for preprocessing is divided into number of sentences, then sentences are further split into a number of words. Further analytics is applied to the smaller token. This is also called as "token analysis" or segmentation. Segmentation is preferred when larger texts are taken for analysis. The process of tokenization is presented in Figure 2.12.

BIG DATA PRE-PROCESSING METHODS

An enhanced ACO algorithm to select features for text categorization and its parallelization.
(Meena MJ et al., 2012)

On the use of MapReduce for imbalanced big data using random forest
(del Río S., 2014)

Large imbalance data classification based on MapReduce
(Park SH et al., 2014)

A parallel algorithm for data cleansing in incomplete information systems using MapReduce
(Chen F et al., 2014)

An extremely imbalanced big data bioinformatics
(Triguero I et al., 2015)

Distributed entropy minimization discretizer for big data analysis
(Ramírez-Gallego S et al., 2015)

An Integrated Data Preprocessing Framework Based on Apache Spark for Fault Diagnosis of Power Grid Equipment
(Shi et al., 2017)

A study on combining dynamic selection and data preprocessing for imbalance learning
(Anandarup Roy et al., 2018)

Online Feature Selection (OFS) with Accelerated Bat Algorithm (ABA) and Ensemble Incremental Deep Multiple Layer Perceptron (EIDMLP)
(Renuka Devi, D., Sasikala 2019)

ACO-PSO-based framework for data classification and preprocessing in big data
(Dubey, A.K et al., 2020)

FIGURE 2.10   Overview of big data preprocessing algorithms.

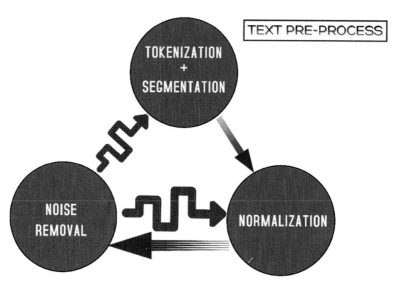

FIGURE 2.11   Text preprocessing method.

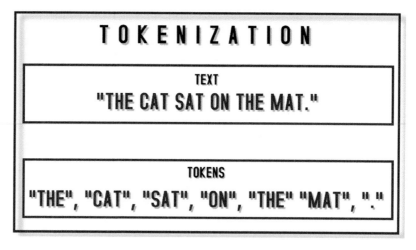

FIGURE 2.12   Tokenization.

Normalization

Converting the texts into a form of uniformity is called "Normalization." Stemming and lemmatization are the basic steps of normalization. Stemming is applied to remove the affixes from the word. For example, if the word is "Travelling," after stemming the word "Travel" is derived. Lemmatization is to form the lemma of the word given.

The other methods are converting into upper/lower case, removal of punctuation/stop words, white space removal and numbers removal, and so on. The following example (Figure 2.13), the raw text forms are normalized into a meaningful form.

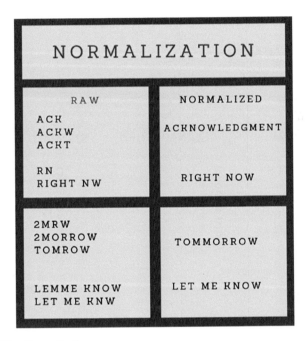

FIGURE 2.13 Normalization.

Noise Removal

Noise removal is the process of removing the unnecessary elements from the given text. This is imperative in text analytics, and problem-specific. When we analyze tweets, it is obvious that the tweeted text may contain hashtags or some special characters. Those characters have to be treated and removed because they may lead to noise.

Let us explain with an example (Figure 2.14), noisy input is converted into stemmed word that still contains the meaningless form. When the noise is removed or cleaned before the stemming process, it will result in a proper lemma form.

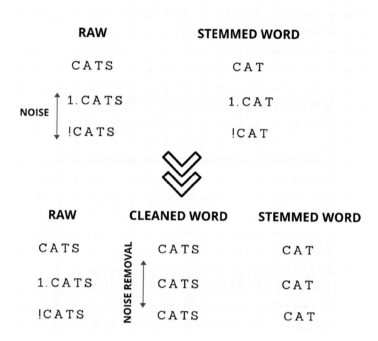

FIGURE 2.14   Noise removal.

2.4.2   *IMAGE PREPROCESSING METHOD*

The objective of image processing is to enhance the image. This is done by removing the distortions and noises. The broad categories of image preprocessing techniques are pixel transformation, geometric transformations, filtering, segmentation, and image restoration.

Pixel Transformation

Pixel transformation changes the brightness of the pixels based on the properties of the image. The output of the transformation depends only on the input. Image transformations include contrast adjustment, color correction, and brightness adjustment.

Enhancing the contrast of the pixels is really significant in the field of medicine, vision, speech recognition, video processing, etc. The types of brightness transformations are brightness corrections and gray-scale transformation. The commonly applied operations are Gamma correction, sigmoid stretching, and histogram equalization.

Gamma Correction

Gamma correction is applied to individual pixels when the objects present in the image are not clear (Figure 2.15). The correction is nonlinear transformation to the pixels present. This is represented by the equation

$$o = \left(\frac{I}{255}\right)^{\gamma} \times 255$$

Original Image **Gamma =2.5**

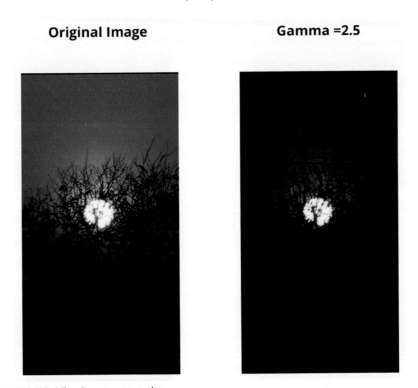

FIGURE 2.15 Gamma correction.

Sigmoid

Sigmoid or logistics function is applied for contrast adjustment (Figure 2.16). This statistical equation is given as:

$$f(x) = \frac{1}{1+e^{-tx}}$$

$$g(x,y) = \frac{1}{1+e^{(c*(th-fs(x,y)))}}$$

The adjusted image is $g(x, y)$, c = contrast factor, th = threshold, and the input image is $fs(x, y)$.

Original Image

Contrast=1

Contrast=2

Contrast=4

FIGURE 2.16   Sigmoid.

When we change the contrast factor, the images change accordingly.

Histogram

A histogram is a common enhancement technique used in most of the applications. The advantage of this method is its compatibility. It can be applied to all types of images and shows enhanced performance. The histogram method employs a nonlinear transformation function to map the intensity of the input and output pixel.

The equalization is applied using a function $P(N)=$ number of pixels with intensity/total number of pixels. The equalization transformation is shown in Figure.2.17.

FIGURE 2.17 Histogram equalization.

KEYWORDS

- machine learning
- preprocessing
- data cleaning
- normalization
- data integration
- data transformation
- data reduction

REFERENCES

Roy, A.; Rafael, M. O.; Cruz, R. S.; George, D. C. Cavalcanti. A Study on Combining Dynamic Selection and Data Preprocessing for Imbalance Learning. *Neurocomputing* **2018**.

Chen, F.; Jiang, L. A Parallel Algorithm for Data Cleansing in Incomplete Information Systems Using Mapreduce. In *10th International Conference on Computational Intelligence and Security (CIS)*; Kunmina: China, 2014; pp 273–277.

del Río, S.; López, V.; Benítez, JM.; Herrera F. On the Use of Mapreduce for Imbalanced Big Data Using Random Forest. *Inf. Sci.* **2014**, *285*, 112–137.

Dubey, A. K.; Kumar, A.; Agrawal, R. An Efficient ACO-PSO-Based Framework for Data Classification and Preprocessing in Big Data. *Evol. Intel.* **2020**.

García, S.; Ramírez-Gallego, S.; Luengo, J. Big Data Preprocessing: Methods and Prospects. *Big Data Anal.* **2016**, *1*, 9. https://doi.org/10.1186/s41044-016-0014-0

Hall, M. A.; Smith, L. A. Practical Feature Subset Selection for Machine Learning, 1998; pp 181–191.

https://machinelearningmastery.com/what-is-imbalanced-classification/

https://medium.com/@limavallantin/dealing-with-data-preprocessing-problems-b9c971b6fb40

https://sci2s.ugr.es/noisydata

https://serokell.io/blog/data-preprocessing

https://towardsdatascience.com/data-preprocessing

https://towardsdatascience.com/having-an-imbalanced-dataset-here-is-how-you-can-solve-it-1640568947eb

https://www.geeksforgeeks.org/data-reduction-in-data-mining/

https://www.kdnuggets.com/2017/12/general-approach-preprocessing-text-data.html

https://www.kdnuggets.com/2019/04/text-preprocessing-nlp-machine-learning.html

https://www.mygreatlearning.com/blog/introduction-to-image-preprocessing/

https://www.tutorialspoint.com/big_data_analytics/big_data_analytics_lifecycle.htm

Tanha, J.; Abdi, Y.; Samadi, N.; Razzaghi, N.; Asadpour, M. Boosting Methods for Multi-Class Imbalanced Data Classification: An Experimental Review. *J. Big Data* **2020**, *7*, 70.

López, V.; Fernández, A.; García, S.; Palade, V.; Herrera F. An Insight into Classification with Imbalanced Data: Empirical Results and Current Trends on Using Data Intrinsic Characteristics. *Inf. Sci.* **2013**, *250*, 113–114.

Meena, M. J.; Chandran, K. R.; Karthik, A.; Samuel, A. V. An Enhanced ACO Algorithm to Select Features for Text Categorization and Its Parallelization. *Expert Syst. Appl.* **2012**, *39* (5), 5861–5871.

Pan, T.; Zhao, J.; Wu, W.; Yang, J. Learning Imbalanced Datasets Based on SMOTE and Gaussian Distribution. *Inf. Sci.* **2020**, *512*, 1214–1233.

Park, SH.; Ha, YG. Large Imbalance Data Classification Based on Mapreduce for Traffic Accident Prediction. In *8th International Conference on Innovative Mobile and Internet Services in Ubiquitous Computing (IMIS)*; Birmingham, 2014; pp 45–49.

Ramírez-Gallego, S.; García,S.; Mourino-Talin, H.; Martínez-Rego, D.; Bolon-Canedo, V.; Alonso-Betanzos, A.; Benitez, JM.; Herrera F. Distributed Entropy Minimization Discretizer for Big Data Analysis under Apache Spark. In *IEEE TrustCom/BigDataSE/ISPA*, Vol. 2; IEEE: USA, 2015; pp 33–40.

Renuka Devi, D.; Sasikala, S. Online Feature Selection (OFS) with Accelerated Bat Algorithm (ABA) and Ensemble Incremental Deep Multiple Layer Perceptron (EIDMLP) for Big Data Streams. *J. Big Data* **2019**, *6*, 103.

Shi, W.; Zhu, Y.; Huang, T. et al. An Integrated Data Preprocessing Framework Based on Apache Spark for Fault Diagnosis of Power Grid Equipment. *J. Sign. Process Syst.* **2017**, *86*, 221–236.

Triguero, I.; del Río, S.; López, V.; Bacardit, J.; Benítez, J. M.; Herrera, F. ROSEFW-RF: The Winner Algorithm for the ECBDL'14 Big Data Competition: An Extremely Imbalanced Big Data Bioinformatics Problem. *Knowl-Based Syst.* **2015**, *87*, 69–79.

Zaki, M. J.; Meira, W. *Data Mining and Analysis: Fundamental Concepts and Algorithms*; Cambridge University Press: New York, 2014.

CHAPTER 3

Feature Selection Methods and Algorithms

ABSTRACT

In this chapter, various feature selection methods, algorithms, and the research problems related to each category are discussed with specific examples. From the diverse sources, data have been gathered, which might contain some anomalies. The voluminous feature sets are combined with pertinent, extraneous, and misleading features during data collection. Feature selection is mainly employed for substantial feature selection with importance. The basic types of feature selection methods used are filter, wrapper, embedded and hybrid. In addition to the basic methods, the online feature selection methods are also highlighted with the use of swarm intelligence based algorithms.

3.1 FEATURE SELECTION METHODS

As we have already discussed in the previous chapters, data volume has been increasing at a rapid pace, thus the numbers of features/attributes are exponentially increased. Number of research interests and approaches have been proposed by scholars and research communities.

From the diverse sources, data has been gathered, which might contain some anomalies. The voluminous feature sets are combined with pertinent, extraneous, and misleading features during data collection. The scalability issue is well addressed by several feature selection (FS) methods with big data technology.

Research Practitioner's Handbook on Big Data Analytics. S. Sasikala, PhD, D. Renuka Devi, & Raghvendra Kumar, PhD (Editor)
© 2023 Apple Academic Press, Inc. Co-published with CRC Press (Taylor & Francis)

FS is mainly employed for substantial FS with importance. The new challenge is added upon the prevailing methods, since the big data challenge is added upon the scenario. The big data FS methods are aimed at not only significant FS, also irrelevant feature exclusion.

The collected data sets are fed into the Machine Learning model for analysis, not all the features contribute to this process. Even though the best features are selected from a huge pool of data, the high dimensionality will generate scalability glitches. Feature exclusion can considerably improve the performance of the model and decrease the time complexity.

The advantages of FS are discussed below:

- Complex models are not desirable for analysis, where simple model is considered for better analytics with significant features.
- Time factor is vital for any data analytics because it is proportionally related to cost factor. Reduced time complexity can lead to faster prediction with affordable cost.
- FS is intended to avoid overfitting problems. The noisy variables having less prediction capability are removed during the FS process, so the generalization of the model can be significantly improved.
- Sometimes, highly correlated features will provide similar information. Removal of those features will enhance the model prediction with less noise factor.

In addition to FS, feature engineering and dimensionality reduction are the terms related to this process. Feature engineering is to create a new set of features from the available features, whereas dimensionality is involved with reducing the number of features or completely transforming the features into low dimension.

3.2 TYPES OF FS

FS are classified into three procedures based on the technique on which features are selected.

They are the following:

1. Filter
2. Wrapper
3. Embedded
4. Hybrid

Filter methods are classifier independent. The wrapper method is feed-back based which works each time the performance of the model is taken into consideration. Embed method employs the combination of filter and wrapper method, and it combines the advantages from both the methods for improved accuracy.

3.2.1 FILTER METHOD

This approach is model independent and classifier independent. The features are selected based on the factor correlation. Correlation exhibits the connectivity between the input and output variables. The features with the highest correlation value are considered as the best feature and chosen for building the model either classification or regression (see Figure 3.1). The advantage of this method is enhanced computation time with tolerant to overfitting problems.

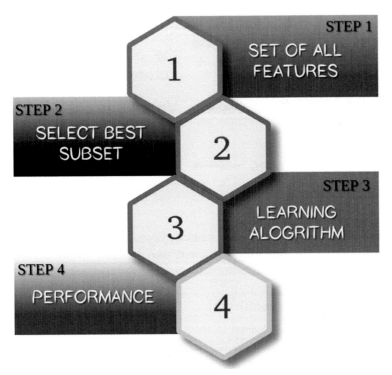

FIGURE 3.1 Filter method.

Filter methods are further split into univariate and multivariate. Univariate is applied on single feature and ranked accordingly. The single feature mapping is applied, and rank is generated individually based on some criteria.

The process is given below:

- Rank the features based on certain criteria.
- Select the highest ranked feature.

The drawback of this method is that it may select a redundant feature, because the correlation between the features is completely ignored. To overcome this type of problems, multivariate is applied on the entire features, considering the features are correlated with each other based on some relevance. The basic filter methods are constant, quasi-constant, and duplicated (Figure 3.2).

BASIC FILTERS

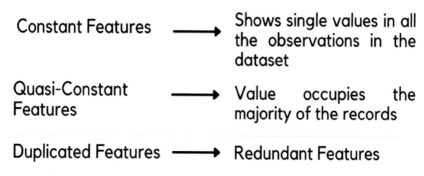

FIGURE 3.2   Basic filters.

Correlation filter methods are, namely, Pearson correlation coefficient, Spearman's rank correlation coefficient, and Kendall's rank correlation coefficient.

Pearson correlation coefficient

This is a known measure used in most of the machine learning models. The linear relationship between the variables is represented between 1 and −1. The positive correlation is represented by the value 1 and negative correlation is represented by −1. If there is no relationship between the variables then the value zero is assigned. Before we apply this method, certain assumptions are established and they are the following:

- The variables are considered to be normally distributed.
- There exists a relation between the variables, represented by straight lines.
- The feature points are distributed around the regression line equally.

The equation is represented below as

$$r_{xy} = \frac{\sum_{i=1}^{n}(x_i - \bar{x})(y_i - \bar{y})}{\sqrt{\sum_{i=1}^{n}(x_i - \bar{x})^2}\sqrt{\sum_{i=1}^{n}(y_i - \bar{y})^2}}$$

r = correlation coefficient

x_i = values of the x-variable in a sample

$\bar{x}$ = mean of the values of the x-variable

y_i = values of the y-variable in a sample

$\bar{y}$ = mean of the values of the y-variable

Spearman's rank correlation coefficient

This method is applied for nonparametric tests, which is used to associate with the degree of correlation. The score of the variables ranges between +1 and -1. It also establishes monotonic associations whether they are linearly correlated or not. The equation is presented below:

$$\rho = 1 - \frac{6\sum d_i^2}{n(n^2 - 1)}$$

ρ = Spearman's rank correlation coefficient

d_i = difference between the two ranks of each observation

n = number of observations

Kendall's rank correlation coefficient

This is the another nonparametric test-based measure, which explores the ordinal relationship between the variables. The measure takes the value between the range +1 and -1. When it is applied on the discrete variables, positive values are given when the rank is similar and negative score is given for dissimilar rank between variables.

The equation is represented below as:

$$\tau = \frac{2(n_c - n_d)}{n(n-1)}$$

where n_c and n_d are concordant and discordant with size n, respectively.

Other methods

Chi-squared and ANOVA

The statistical methods like ANOVA and chi-square are generally used to test for the categorical variables. The correlation score is generated between categorical variables with the specification that it should be only suitable for nonnegative variables.

Similar to this method, ANOVA scores the dependency between the variables, with assumption that they are normally distributed and establishes the linearity between the variables.

3.2.2   WRAPPER METHOD

Wrapper-based approach is a kind of feedback analysis, where FS is applied to the subset of features, and model is trained and accuracy is measured (see Figure 3.3). Based on the previous results, we tend to add the feature or remove it from the datasets. This process is iterative to achieve a promising accuracy with the selected features.

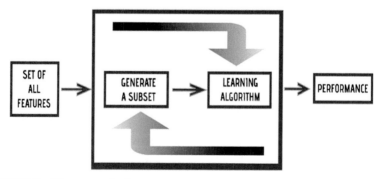

FIGURE 3.3   Wrapper method.

The generalized process is given below:

- Employ search method for a subset
  - ○ Forward — Select the features by adding one after the other
  - ○ Backward — From the given features set, remove one by one
  - ○ Exhaustive — Combination of all methods
  - ○ Bi-directional — Employs both forward and backward method
- Develop a Machine Learning Model—based on the subsets formed.
- Evaluate a model.
- Iterate the above process until the condition is arrived. The iterative process must be stopped when the performance of the model degrades/performance of the model increases, and the selected features are substantial features are selected for model building.

3.2.3 EMBEDDED METHOD

It takes and combines all the advantages of both the filter and wrapper methods.

SELECTING BEST SUBSET

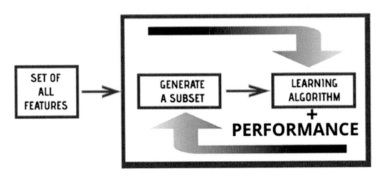

FIGURE 3.4 Embedded method.

The generalized step of this method is given (Figure 3.4) below:

- Form a subset of features.
- Create and train the model.

- Derive the feature importance between the variables/score.
- Remove the nonredundant features.

The most commonly used embedded methods are regularization and tree-based methods. L1, L2, and L1/L2 are the regularization methods used. Tree-based method uses random forest for FS, where information gain/entropy is calculated for FS decision.

The major advantage of this method is that,

- faster,
- accurate, and
- prone to overfitting.

3.2.4 HYBRID

This method is the combination of taking advantages from all the previous methods for the FS process. The feedback from the previous learning experience coupled with the ranking of the features.

Advantages

- fast
- accurate
- scalability
- enhanced computational complexity

3.3 ONLINE FS METHODS

Streaming FS is called "Online Feature Selection (OFS)," involved with real-time analytics. This is vital for stream data mining. The FS process is to select the significant features from streams of incoming data then the machine learning model is formed based on the selected features (Al Nuaimi and Masud, 2020).

Most of the conventional systems access all the features at once (Single Batch) before the analytics process. When we consider the real-time applications, not all the features are available at the beginning, because of the streaming nature the instances may arrive at different intervals of time. To address this issue, there should be a mechanism required for updating the instances and the induction model. OFS processes the instances on the fly.

The streaming FS challenges are (Wang et al., 2014)

- When all the instances are available at the beginning of the model building.
- When instances arrive at an irregular interval—partial learning of instances.
- When instances are streaming nature.

Most of the FS methods are based on batch processing mode (offline mode). OFS methods are used to process the features online as they arrive (Devi et al., 2019). In data stream mining (Stefanowski et al., 2014), this approach is used to select the significant features to create a model with enhanced prediction accuracy.

Several approaches have been suggested in recent literatures for streaming features selection, such as maintaining the constant number of selected features even though the features arrive one at a time (li et al., 2017).

The streaming feature grouping approach is proposed for selecting the small number of significant features from Big Data (Al Nuaimi and Masud, 2020). The 5Vs of big data characteristics have presented challenges of streaming FS.

The social network-based OFS approach is applied for FS in processing the social text streams (Tommasel and Godoy, 2018). The social data is highly complex and voluminous. The classification of text streams is really a challenging task here.

3.4 SWARM INTELLIGENCE IN BIG DATA ANALYTICS

The metaheuristic methods are generally used for large-scale optimization problems. Evolutionary computation methods/nature-inspired approaches are employed for Big Data analytics problems. Compared with conventional algorithms, swarm intelligence (SI)-based methods provide enhanced and optimal FS (Sasikala and Devi, 2017).

The advantage of this method is search capability and feature space exploration with randomness. These methods are well suited for optimal FS. Some of the methods (Rostami et al., 2021) are Genetic Algorithms (GAs), Ant Colony Optimization (ACO), Particle Swarm Optimization (PSO), Ant Bee Colony (ABC), and Bat Algorithm (BA). Some of the algorithms are detailed in this section.

The SI is the algorithm derived from the collective population of individuals, also called the nature-inspired techniques. The key aspect of any nature-based approaches is to provide optimal solution by exploring and searching the solution space. The key element is population, which consists of a group of possible solutions spread across the solution space. In contrast with the single-point algorithm, this is entirely a population-based algorithm.

The conventional approaches are primarily focused upon handling the collected data with the predefined statistical methods (Yang et al., 2020). When the perception of big data came into effect because of huge data accumulations, thus complexity of the conventional analytics approaches has been increased. Therefore, novel optimization algorithms are employed in the place of conventional types. The optimization algorithms are generally preferred because of its scalability and the efficacy to encounter the other V's of Big Data characteristics (Cheng et al., 2013).

There is a paradigm shift from the conventional approach to data-driven approach. Most of the present-day scenarios in the domain of data sciences are very well handled by optimization problems. The conventional types force the problem to be presented in the continuous and differentiable methods, which is tuned into a complex procedure when the scalability is higher. So, it is always beneficial to use the optimization algorithms.

There are two kinds of heuristics approaches. Simple heuristics approach is dependent on a specific problem (problem specific) and metaheuristic approach is independent of problem, having the ability to handle large-scale problems, providing multiple solutions. SI is a kind of metaheuristics algorithm.

3.4.1 ELEMENTS OF SI

The basic (Kennedy, 1995) element of SI is "Swarms". The general characteristics are shown in Figure. 3.5.

SI algorithms consist of a given population of candidate solutions that are grouped together. They are randomly placed in the population, not in any particular order. Every individual element is considered as the potential solution that is to be optimized for every iteration. The iteration is the number of steps required to repeat the process, so that for every iteration there is a tendency of generating good solutions. This process is repeated until it reaches the stopping criterion or good solution. The solutions are guided at every iteration to move forward for betterment.

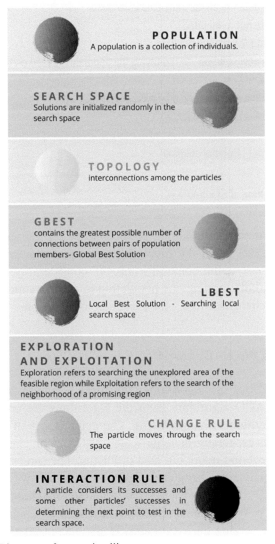

FIGURE 3.5   Elements of swarm intelligence.

The outline of the algorithm is presented in Figure.3.6.

Generalized Algorithm

Step 1: Initialize the random Population -Generate random Solution
Step 2: Evaluate the all the initialized individuals
Step 3: while not terminated
    Calculate the fitness function
    Form a new solution , by modifying the population based on the local and global best
    Update the new solution if best otherwise discard
    Update the iteration counter
    end
Step 4: Output : Optimal solution

FIGURE 3.6   Outline of SI algorithm.

As already mentioned, the algorithm iterates for the specific number of times. For every time, the fitness function is calculated for a time t. Over the time $t + 1$, the solution improves considerably. The updating process is a kind of mapping one population to another represented by $f()$. Population(p_t) changes into p_{t+1} for every step of iteration.

3.4.2 SI AND BIG DATA ANALYTICS

Among the various SI algorithms, the most common ones are PSO algorithm and ACO algorithm. Both the algorithms are commonly used for single or multiobjective problems. The applications of SI algorithms are widespread, especially in large-scale clustering or classification, FS problems, and solving many big data problems. The sandwich of both SI and Big data Analytics (Cheng et al., 2016) is given below (Figure 3.7).

Big data analytics is envisioned for large data handling with reduced time complexity. When the complexity of the data and pattern increases, the objective of solving those problems is also increased along with time complexity.

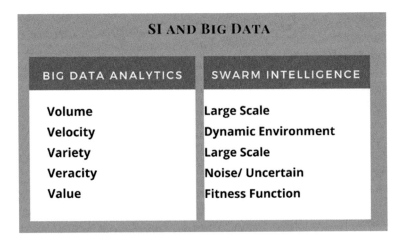

FIGURE 3.7 SI and big data analytics.

The objective of optimization is to provide multiple solutions within a given time period. The solutions can be several or local solutions. These kinds of multiobjective, scalability problems are well handled by SI algorithms merged with big data techniques. The following section addresses this issue. The preview of this is depicted in Figure. 3.8.

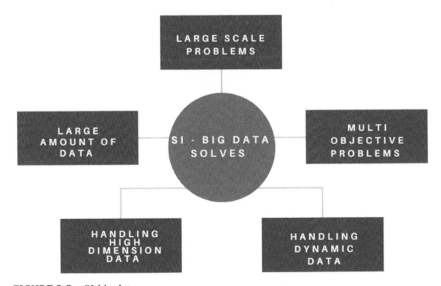

FIGURE 3.8 SI-big data.

Handling large data requires a quick scan to the entire search space. This is a population-based sampling technique where the entire solution space is searched and explored for optimal solution. As per the rule of metaheuristic concepts, the sampling is done from the massive population.

The representative solution is obtained from the large population when we use SI-based strategy. Thus, we could provide the best solution for large-scale problems. When this approach is coupled with MapReduce, the scalability and time complexity problems are being solved in a more competent way.

Even though the optimization performs well (Cheng et al., 2013)sometimes its performance may deter when the dimension of the population space is increased exponentially. Thus, efficient search strategies must be devised to explore all possible solutions.

Dynamic data changes at a rapid rate. Most social media data are dynamic in nature, such as twitter, facebook, and Instagram. The usage of the Internet has been increased at a rapid rate, thus data rate is escalating at a rapid speed with volume.

This problem is a kind of nonstationary. The real-time streaming big data analytics are used to solve this kind of issue. SI algorithms solve both stationary and nonstationary problems. The various uncertainties that might arise in optimization problem are discussed below:

- Handle noisy data and computing fitness function for the same.
- Optimization might deviate from the optimal point of solution.
- The approximation is done sometimes. The fitness functions suffer with approximation due to external parameters.
- Difficult to track the optimum when it changes over the particular point.

The uncertainties can be challenged by the SI—multiobjective optimization algorithms. The functionalities are given below:

- To reduce the distance between nondominated to the optimal solution.
- Solutions should be distributed uniformly.
- Series of values should be enclosed by nondominated solutions.

The various SI-based algorithms are presented below:

FIGURE 3.9 SI algorithms.

3.4.3 APPLICATIONS

The applications of SI-based algorithms across the verticals, social network analysis, scheduling, gene sequencing, resource allocation, facial recognition, and image segmentation.

Social network analysis is primarily involved in connecting the people across networks. To explore connected community patterns and preference, this kind of analytics plays a vital role. The community detection algorithms have been proposed based on SI algorithms (Lyu et al., 2019) to explore the community structures in a network of complex structure.

In addition to the above-mentioned application, it extends to disperse power/energy systems and consumer-demand analysis. Biological gene sequencing problems are very well addressed by SI. Protein structure design is considered as an optimization problem. Cloud service resource allocation (Cheng et al., 2017) problem is effectively handled by this approach. Natural Language processing (Abualigah et al., 2018) applications have also been proposed.

3.5   PARTICLE SWARM OPTIMIZATION

Particle swarm optimization (PSO) is a stochastic, metaheuristic population-based algorithm developed by Dr. Kennedy (Kennedy, 1995). The social behavior of birds or fish schooling was great inspiration for the development of this algorithm. The problem-solving approach is inspired from the problem-solving nature of birds when they are in search of food.

The social behaviors are impersonated by PSO for finding optimal solutions. The initial version of the approach has been changed considerably over the year to accommodate recent research challenges and variations in the basic weight adjustment, inertia, velocity, and other hybrid methods are proposed.

Functionality and components of Algorithm

The algorithm starts with searching for a particle or a solution, which is called "swarms". As the general rule of any population-based algorithm is searching and moving forward for the better solution for each iteration. The fitness function is applied on every particle.

The two terms are important here: one is local best (lbest) and global best (gbest). The lbest and gbest are evaluated for each iteration, and the particle is attracted toward the global best position (Zhang et al., 2014). The process flow of the algorithm is presented in Figure.3.10.

The algorithms start with initializing the particle along with velocity and position. The fitness function is calculated for each randomly distributed particle. Gbest and lbest are interchangeably updated according to fitness value and swapped with the positions. At the end of each iteration, the position is updated along with the velocity.

The objective function is either to maximize or minimize to reach the optimum. Wide range of optimization methods are available, but they are problem dependent. Depending upon the features count they might vary and applied accordingly. To find the best fit, the optimization method is chosen accordingly.

3.6   BAT ALGORITHM

Bat algorithm (BA) is one of the optimization algorithms based on the echolocation characteristics of the bats. All the nature-inspired algorithms try to mimic the functionality of nature. Sane way, how bat emits sound and recognizes the prey in the path and acts accordingly. Bats when they fly, emit a sound pulse. The echo is sensed by bats when the prey or object is recognized in the path. In this way, the identification is made even in the complete darkness. The same functionality is stimulated by the algorithm to achieve optimal solution.

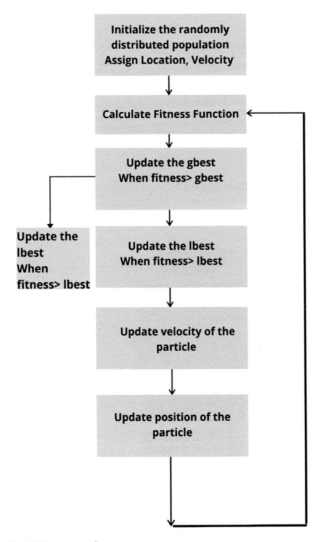

FIGURE 3.10 PSO process flow.

The outline of the algorithm is given below:

1. Bat discover food and prey using echolocation.
2. Every bat has velocity ve_i, with a feature position fp_i with $freq_{min}$ fixed frequency, λ varying wavelength and A_0 loudness. The pulse emission rate varies between $r \in [0,1]$. The wavelength is modified accordingly.
3. The loudness is from A_0 to A_{min}.

Based on the steps explained, the algorithm generates possible optimal solutions by exploring the search spaces. The frequency ranges between $freq_{min}$, $freq_{max}$ corresponds to the wavelength $[\lambda_{min}, \lambda_{max}]$. The velocity and frequency are updated by the following equations (Akhtar et al., 2012),

$$freq_i = freq_{min} + \left(freq_{max} - freq_{min} \right) \beta \qquad (3.1)$$

$$ve_i^t = ve_i^{t-1} + \left(fp_i^t - fp_* \right) freq_i \qquad (3.2)$$

$$fp_i^t = fp_i^{t-1} + ve_i^t \qquad (3.3)$$

$\beta \in [0,1] =$ random vector drawn from a uniform distribution

The flow of the process is presented in Figure 3.11:

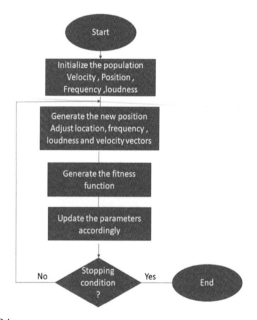

FIGURE 3.11   BA—process.

3.7 GENETIC ALGORITHMS

Genetic algorithm (GA) is based on a mathematical model (Holland et al., 1975;Goldberg, 1989) of optimization. This is a heuristic-based optimization approach, where elements of the population try to provide enhanced solutions. The algorithm mainly consists of three sections. The outline of the algorithm is presented below:

Step 1: Solutions are represented by a string or sequence of real values. The vector is formed with all these elements together.

Step 2: For every iteration, solutions are modified by means of mutation or crossover mechanism, at a single point or multiple points. Crossover is applied to the parent solution to produce the new solution. But mutation is applied for single or multiple points of solution.

Step 3: The objective function, that is, fitness function is evaluated for each solution. Higher value of the function is preferred for maximization problems. Best solutions are compared and chosen at every iteration. The process of the algorithm is presented in Figure 3.12.

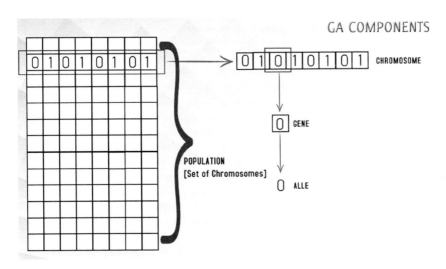

FIGURE 3.12 GA components.

The basic components of GA (Figure 3.6) are the following:

Population	The given set of possible solutions
Chromosomes	A single individual possible solution
Gene	Is the position of the element
Allele	Is the gene value

The above-mentioned procedure is the basics for all GA algorithms. All the biological and genetics-based approaches are based on GA only. When compared with general biological science, genes evolve, mutate, and adapt to the environment. The state diagram is depicted in Figure 3.13.

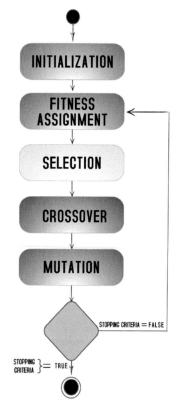

FIGURE 3.13   State diagram.

Figure 3.13 outlines the entire flow of the process. The process is iterated until the stopping condition is reached. The crossover and mutation are done for swapping and forming a new best solution. The objective function is used for evaluation.

The first step is to form the population. To enhance the diversity of the population, the population is randomly distributed. Fitness function is applied upon each individual element of the population. Once the function lists the value, the element with highest score is considered for selection. The commonly used fitness methods are rank-based assessment.

After the selection, the solutions with the highest value are combined and proceeded for the next iteration. When the number of selected solutions reaches half of the total number of given solutions, the crossover mechanism is applied on them to generate new solutions.

This process of combining features, for example, if two selected solutions are combined to form double the features of offsprings to form a new solution. The uniform crossover method checks upon whether the featured are formed from a single parent or another one.

The disadvantage of crossover method is that sometimes it might generate the solution very much similar to that of a parent causing low diversity of the population. To combat this issue, mutation operator is applied (Figure 3.14). Thus, the entire process of fitness evaluation, selection, crossover, and mutation is repeated for a number of times until we achieve the optimal solution.

3.8 ANT COLONY OPTIMIZATION

The ant colony optimization (ACO) methodology was devised by Marco Dorigo in 1990. The ant's nature-inspired algorithm used the food-seeking behavior of the ants for simulation of the algorithm. The seeking nature and finding and gathering aspect of the ants were taken for research. First, this methodology was used to solve the traveling salesman problem. Later, it extended its applications on various research areas, particularly hard optimization problems.

Ants are community-dependant and prefer to live in a group than an individual. The interaction between them is possible through sense, sound, and pheromone. The pheromone is a chemical secreted by the ants during communication and interaction among them. The trails formed by them gives a clue to the rest of ants to follow the path. Normally they are sensed by others to follow. This phenomenon is stimulated for real-time applications as well.

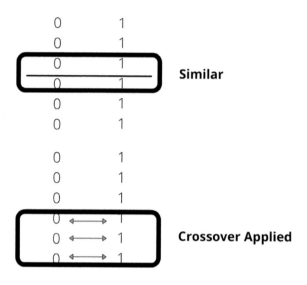

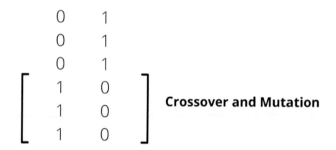

FIGURE 3.14   Crossover and mutation.

The foundations of this methodology are the following:

- To follow and perceive the path—to search for food.
- Multiple paths are found.
- The selected path should be shortest one.

To start with, ants generally move toward a path in search of food in a random manner. The search is random, because all the possibility of reaching the food is explored and found. The path is established like a line between their location and source of food.

Every time when the ants move, they leave the pheromone deposit on the path based on the food quality and amount. Thus, the trails paths are selected based on the probability of getting the food source. The other factors like concentration of pheromone, the shortest distance is also taken for consideration.

The inferences from Figure 3.15 are discussed below.

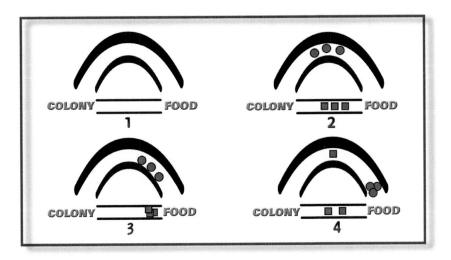

FIGURE 3.15   ACO—path.

Consider the first one, there was no pheromone on the path between colony and food source. This is the initial stage. The probability of the pheromone is residual in this case. The path is not explored and searched.

The second stage, where food source is available, and the path is to be explored. The two paths curved and straight ones are considered for the search. Equal probability of searching for food is done. It is found that the longer path is curved one compared to the straight path that is shorter.

In the next stage, the selection of the path is done with a probability function. The shortest path is considered to be the best one, because of distance and also pheromone concentration. The selection of the shortest path is finalized with high probability value.

At the final stage, when more ants do follow the shortest path because of high concentration of pheromone and the probability of choosing the longest path is discarded completely. Thus, the entire colony moves toward the optimal shortest and concentrated path for food.

The process flow is shown in Figure 3.16.

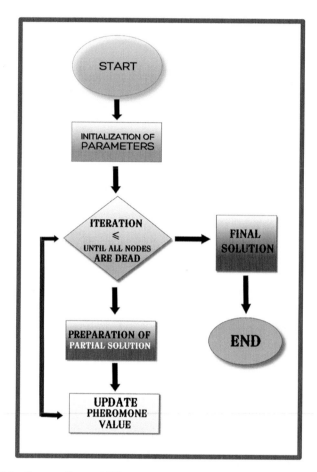

FIGURE 3.16 Process flow—ACO.

The process flow diagram shows how the optimal solution is arrived at. Two factors are considered here:

- **The updating of pheromone value**—The value is updated at every point of movement from one location to another. The ant density and quantity are changed accordingly.
- **Cycle of the process**—The updation is done for all ants when they complete their travel.

Algorithm

- **Initialize** Trail
- **Do White** (Stopping Criteria Not Satisfied) – Cycle Loop
  - **Do Until** (Each Ant Completes a Tour) – Tour Loop
  - Local Trail Update
  - **End Do**
  - Analyze Tours
  - Global Trail Update
- **End Do**

The optimized path arrives when the pheromone is updated with transition probability in the search space by a simple mathematical formula. The fitness function is measured and global fitness value is generated for ants very similar way how the general optimization algorithm works.

The search is iterative and continuous, checking the fitness score. If the fitness score is high, move the ants to that location else start the new search in the random space for discovering a new path. The search is local and global based. It is represented in the following equation:

$$P_k(t) = \frac{t_k(t)}{\sum_{j=1}^{n} t_j(t)}$$

P is the probability function. For a region k, n is the total number of ants and r is the rate of evaporation. t is the total pheromone at a region k.

The path of the ants is depicted as a graph of paths (Figure 3.17). The graph is made up of edges, nodes/vertices, and paths. The source and nest of the ants are nodes. The path connects the food source and ants. Path traversal is implemented through a weighted graph. The weights are nothing but the pheromone scores, associated with every path of the traversal.

The outline of various optimization problem where ACO is used are presented in Figure 3.18.

Applications

- Telecommunication networks
- Cloud-load balancing
- Genomics

- Routing and scheduling
- Dynamic applications
- Large scale FS problems

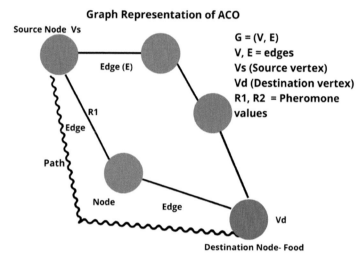

Graph Representation of ACO

Source Node Vs

Edge (E)

G = (V, E)
V, E = edges
Vs (Source vertex)
Vd (Destination vertex)
R1, R2 = Pheromone values

R1

Edge

Path

Node   Edge

Vd

Destination Node- Food

FIGURE 3.17   Graph—ACO.

1. Stochastic vehicle routing problem (SVRP)
2. Vehicle routing problem with pick-up and delivery (VRPPD)
3. Group-shop scheduling problem (GSP)
4. Nursing time distribution scheduling problem
5. Permutation flow shop problem (PFSP)
6. Frequency assignment problem
7. Redundancy allocation problem
8. Traveling salesman problem (TSP)

FIGURE 3.18   ACO optimization methods.

3.9   ARTIFICIAL BEE COLONY ALGORITHM

Artificial bee colony (ABC) is one of the metaheuristic methods devised by Karaboga 2014) for various optimization problems. The life of bees is stimulated by this algorithm. The components of these algorithms (Tereshko and Loengarov, 2005) are discussed below:

- Foraging BEES—They search for good sources of food.
    ○ Employed
    ○ Unemployed—Onlookers, Scout
- Food Sources—Checks nearer to hives
- Behavior
    ○ Self-organized
    ○ Collective
- Feedback
    ○ Positive—To check for good sources and accept it
    ○ Negative—Reject the sources of food

The swarm of bees is assigned a task for searching food (Figure 3.19). They find the sources by means of cooperation between them. The intelligent behavior of the bees is the fundamental approach here. The bees that are employed searching food try to save the location in memory, and later it is shared to onlooker bees.

ABC- FOOD SEARCH

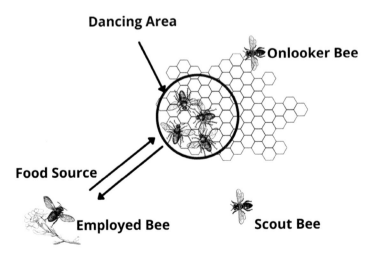

FIGURE 3.19 Food search—ABC.

The duty of onlooker bees is to select the good food from the list of information shared by employed bees. The fitness function or objective function is evaluated for the different sources of food found by the employed bees.

Higher the fitness value, the source is taken for consideration, thus avoiding the low-quality foods. When the employed bees are exhausted in search for food, they are turned into scout bees. The scout bees are hired for searching new sources, when found food sources of low quality. The location where the optimal source is found, it is considered as the possible solution.

The objective function is evaluated and iterated for a number of times to achieve an optimal solution. The initial vectors are assigned with values that are changed accordingly when they move toward the optimal solution by searching mechanisms thus avoiding the poor solutions. The total number of employed bees is equal to the number of solutions.

The general scheme of the ABC algorithm is as follows:

ABC - ALGORITHM

STEP 1: Initialization Phase
STEP 2:
REPEAT
      Employed Bees Phase
      Onlooker Bees Phase
      Scout Bees Phase
      Memorize the best solution achieved so far
UNTIL (Cycle=Maximum Cycle Number or a Maximum CPU time)

STEP 3:
      OUT PUT - OPTIMAL SOLUTION

FIGURE 3.20   ABC algorithm.

Applications

- Large-scale clustering problems.
- Training feedforward neural networks.
- Sensor-based classification.
- Electrical engineering optimization problems.
- Power flow controller—Distributed energy systems.
- Multiobjective optimization problem.
- Load balancing.
- FS.

3.10   CUCKOO SEARCH ALGORITHM

The cuckoo search algorithm is a metaheuristic algorithm (Yang, 2009; Yang and Deb, 2014) used for resolving optimization problems (Joshi et al., 2017). This is inspired by the cuckoo and entirely portrays and simulates the characteristics of cuckoo bird. The intelligent aspect of cuckoo is to lay eggs in host bird's nest, and remove the other eggs to upsurge the hatching.

Parameter tuning is the key aspect of the algorithm. Certain hyperparameters are tuned in such a way that it would enhance the algorithm's efficacy. The cuckoo search for food is supported by a random walk (levy flights).

The random walk is represented by the graph of Markov Chain. The search position is moved to a position based on the current location and transition probability. The search capability further can be expanded by levy flights. Levy flights are the way they search for a food in a random fashion. The movement or the path they take depends on the current position and probability of steps (probability distribution function).

Each cuckoo is considered as the potential solution for the given problem. At the end of each iteration, the improved solution is generated and is substituted with the formerly generated solution. So, when the iterations are completed, we accomplish a good solution for the complex problem. Certain considerations are outlined in this algorithm, which are as follows:

- Nests have one egg each.
- For complex problem, more than one egg is placed in the nest.
- Nests are randomly chosen and eggs are placed (cuckoo's egg as new solution)
- No of eggs = No of potential solutions.

- Fitness function is applied, the nest with the highest fitness will be given a priority.
- The total number of the host's nest is a constant.
- The probability of host's bird to identify the misplaced egg will be from zero to one.
- Based on the probability, the host either will remove the egg or nest.

The applications of cuckoo search algorithm in various domains are listed below:

- Np hard problems.
- FS.
- Neural Network—Training.
- Design optimization.
- Speech analytics.
- Engineering Applications.
- Image Processing.
- Health care (ECG).
- Wireless Networks.
- Robotic Path.
- The generalized algorithm is given in Figure 3.21.

3.11 FIREFLY ALGORITHM

Firefly algorithm is another SI-based algorithm, inspired on the fireflies' characteristics. The mode of communication between them is signaled by flashes. The primary objective of the fireflies is that to flash and attract each other (Xin-She Yang, 2009).

The fundamental characteristics are the following:

- Fireflies attract each other irrespective of their sex (considered all flies are unisex).
- The attractiveness is proportional to the brightness (shorter the distance, higher the brightness)
- The attractiveness is inversely proportional to the distance.
- Fireflies are attracted only for higher brightness.
- Fitness function is applied for calculating the brightness level.
- For optimization problems, brightness is proportional to the fitness function.

CUCKOO SEARCH ALGORITHM

Step1 : Randomly initialize the population of host nests(n)

Step 2: Place the egg in random nest say k. Random walk (levy flights to place the egg)Cuckoo's egg is very similar to the host egg.

step3: Generate the fitness function

Step 4: Compare the fitness function of the cuckoo with the host

Step 5: if fitness(cuckoo) > fitness(host)
          replace the egg
          else
          the nest is abandoned

Step 6: Iterate the step 2 to 5 , until the condition is met(the optimized solution is obtained)

FIGURE 3.21   Cuckoo search algorithm.

The generalized algorithm is specified in Figure 3.22. The standard algorithm begins with the initialization step, where the flies are randomly populated. Each fly is associated with light intensity. The intensity of the light gives the level of brightness and the flies are fascinated by the bright light.

The movement of the flies is based on this intensity level and the positions are updated. The objective function (fitness function) is applied to each fly. The local best solution is compared with global best and replaced according. This process is iterative until the maximum generation is achieved.

FIREFLY ALGORITHM

Step1 : Initialization

Randomly initialize the population of fireflies.
The Light intensity of each firefly is measured and updated accordingly.

Step 2: Brightness (Attractiveness) and Distance

Attractiveness is measured based on the brightness.
The objective function is applied and evaluated for the same.

Step 3: Fitness Function

Generate the fitness function

Step 4: Optimal Solution (Next Generation)

Find the current best solution and replace .
Movement of all Fireflies towards the better solution

Step 5: Termination

Iterate the step 2 to 4, until it reaches the maximum generation and terminate

FIGURE 3.22 Firefly algorithm.

The algorithm is very well suited for parallel implementation. Many research studies have proposed parameter tuning because of slower convergence and local optimum problems. The basic algorithm is primarily projected for noncontinuous problems; hence the parameters should be fine-tuned for a continuous problem.

The adjustments of parameters are suggested (Khan et al., 2016) below:

- Parameter Changes
  ○ Change the parameters updating mechanism of the algorithm is same.
- Mechanism Changes
  ○ Propose novel update mechanism/function.
  ○ Use hybrid methods, mutation operators.
- Search Space Changes
  ○ Use probability distribution function.

The Firefly algorithm is simple and easy to implement. The various applications are listed below:

- FS.
- Big data optimization.
- Engineering.
- Text classification.
- Wireless sensor networks.

3.12 GREY WOLF OPTIMIZATION ALGORITHM

Grey wolf optimization (GWO) is one of the metaheuristics methods used for resolving optimization problems. The grey wolf procedure is extensively used for major optimization problems for its added advantages. The algorithm is extensively used for large-scale problems (Faris et al., 2018) and flexibility in handling complex problems.

The optimal convergence is achieved through sustaining the proper balance between exploration and exploitation. Grey wolves are considered as topmost predators and occupy the highest level in the hierarchy in the food chain and they live as a group. The GWO stimulates the hunting behavior of them.

The classifications of grey wolves are alpha, omega, beta, and delta. The alpha wolves are mainly involved in decision-making, search, and hunting operations. Others merely follow them. The group hunting behavior is characterized as the following:

- Search for prey and track.
- Encircle the prey, and stop the movement of the prey.
- Finally, attack mode.

The replications of the same behavior and hunting characteristics are depicted as mathematical model. As mentioned above, the alpha wolves are decision maker, they are considered as the best solution to the optimization problem. The optimization is directed by alpha, omega, beta, and delta parameters. The positions of the grey wolves are updated at each iteration, thus an optimal solution is reached.

The outline of the algorithm is presented in Figure 3.23. The standard algorithm gives the best solution over the number of iterations. However, the parameters can be enhanced and fine-tuned. The encircling mechanism leverages the functionality that can be also be extended for high dimension space. Also, there is a possibility of integrating mutation and evolutionary operators.

Applications
- Global optimization
- FS
- Engineering applications
- Prediction modeling
- Image segmentation
- Clustering applications

3.13 DRAGONFLY ALGORITHM

Dragonfly algorithm (DA) is SI-based metaheuristic optimization algorithm (Mirjalili, 2016) stimulated by the swarming behavior of dragonflies. Nature-enthused algorithm is used for solving variety of optimization problems (Meraihi et al., 2020). Dragonflies are a smaller insect that hunts for prey like ants and bees.

The algorithm is based on the static and dynamic swarming features of the flies (Baiche K et al., 2019). The static and dynamic swarms are the phase where exploration and exploitation of swarm begins. The static swarm behavior is to fly in one direction and divert from enemies. The dynamic nature of flies to form a small group and fly over an area in search of food (Figure 3.24).

GREYWOLF ALGORITHM

Step1 : Initialization

Initialize the the population of grey wolves.
t=Current Iteration
Initialize A,C (Coefficient Vectors)

Step 2: Generate Fitness Function

Calculate the fitness function for each search agent. where
Alpha = Best Search Agent
Beta = Second Best Agent
Delta = Third Best Agent

Step 3: Update Function

Update the position of the search Agent with the best one.
Calculate the objective function
Update the fitness value for all

Step 4: Termination

Iterate the step 2 to 3 , until it reaches the maximum number of iterations.

FIGURE 3.23   Grey wolf algorithm.

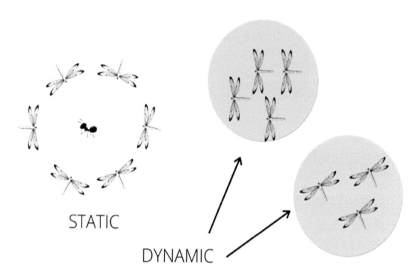

FIGURE 3.24   Swarming behavior.

The basic ideologies of these algorithms (Figure 3.25) are presented below:

- Separation—Collision is avoided.
- Alignment—Fly at their own speed in line with others.
- Cohesion—Tendency to move toward the center point of swarm.
- Attraction—Toward the food source.
- Distraction—Divert from opponents.

The algorithm is presented below:

1. Initialize the population
2. Calculate the fitness function for each fly
3. Update the source of food, other parameters like radius
4. If neighboring fly exists, then update the velocity and position of the fly otherwise update the velocity
5. Update the new positions and boundaries
6. Repeat the steps 2 to 5 until the maximum iteration is reached.

The applications are extensive across verticals. They are FS, resource allocation, routing problems, ANN-training, image segmentation, optimization, MLP networks, optical filters, medical image processing, intrusion detection, energy optimization, RFID networks, and healthcare sector.

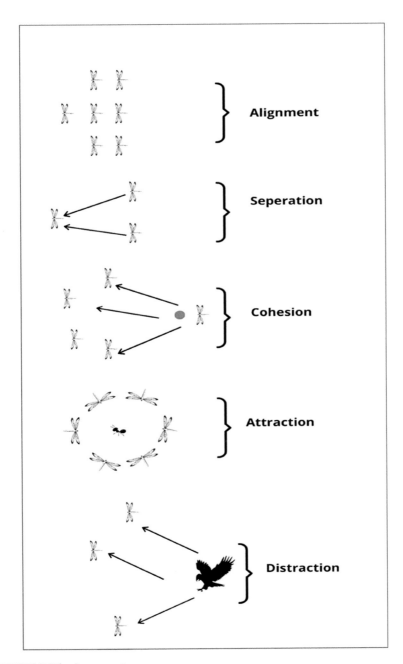

FIGURE 3.25 Swarm patterns.

3.14 WHALE OPTIMIZATION ALGORITHM

Whale optimization algorithm (WOA) (Mafarja and Mirjalili, 2017) is the new algorithm developed under the group of metaheuristics optimization. The algorithm is grounded on the bubble net hunting technique of whale's hunting behavior.

It is widely used for engineering problems and other applications because of its simplicity, right balance between exploration and exploitation, less hyper parameters, robustness, and the convergence of the algorithm is fast (Rana et al., 2020).

The basic components of the algorithm are the following:

- **Encircle:** Prey is encircled when the location is found by the whales. The search space is unknown at the beginning of the search. The assumption is made in such a way that the present solution is considered as "prey" and the position of other agents is updated accordingly.
- **Exploitation:** Bubble net attacking method—The position of the whale follows the spiral path to attack the prey.
- Exploration: Search for prey—The search agents are updated with new positions of best search agent.

The generalized procedure is given below:

1. Initialize the whale positions, and they are randomly set.
2. Generate an objective (fitness function) for each candidate.
3. The probability is found and the positions are updated based on that value. (helix-shaped movement).
4. One component of each whale is changed with a probability.
5. Repeat steps 2 to 4, until a termination condition is met.
6. Report the best solution.

Applications

- Classical Engineering problems—Bar design, beam design, and so on,
- Power systems
- Robotics
- Optimization techniques
- Navigation radar
- Machine learning

- Neural network
- Civil engineering

KEYWORDS

- **swarm intelligence**
- **big data**
- **particle swarm optimization**
- **ant colony optimization**
- **genetic algorithm**
- **feature selection**
- **bat algorithm**

REFERENCES

Abualigah, L. M.; Khader, A. T.; Hanandeh, E. S. A New Feature Selection Method to Improve the Document Clustering Using Particle Swarm Optimization Algorithm. *J. Comput. Sci.* **2018,** *25,* 456–466.

Akhtar, S.; Ahmad, A. R.; Abdel-Rahman, E. M. A Metaheuristic Bat-Inspired Algorithm for Full Body Human Pose Estimation. In *Ninth Conference on Computer and Robot Vision* 2012; pp 369–375.

Al Nuaimi, N.; Masud, M. M. Online Streaming Feature Selection with Incremental Feature Grouping. *Wiley Interdiscipl. Rev. Data Mining Knowl. Discov.* **2020,** *10* (4), e1364.

Baiche, K.; Meraihi, Y.; Hina MD.; Ramdane-Cherif, A.; Mahseur M. Solving Graph Coloring Problem Using an Enhanced Binary Dragonfly Algorithm. *Int. J. Swarm Intell. Res. (IJSIR)* **2019,** *10* (3) 23–45.

Cheng, S. et al. Cloud Service Resource Allocation with Particle Swarm Optimization Algorithm. In *BIC-TA 2017. CCIS*; He, C., Mo, H., Pan, L., Zhao, Y., Eds., Vol. 791; Springer, Singapore, 2017; pp 523–532.

Cheng, S.; Shi, Y.; Qin, Q.; Bai, R. Swarm intelligence in Big Data Analytics. In *IDEAL. LNCS*; Yin, H. et al., Eds., Vol. 8206; Springer, Heidelberg, 2013; pp 417–426.

Cheng, S.; Zhang, Q.; Qin, Q. Big Data Analytics with Swarm Intelligence. *Ind. Manage. Data Syst.* **2016.**

Eberhart; Russell; Kennedy, J. Particle Swarm Optimization. In *Proceedings of the IEEE International Conference on Neural Networks*; Citeseer, 1995.

Faris, H.; Aljarah, I.; Al-Betar, M.A. Grey Wolf Optimizer: A Review of Recent Variants and Applications. *Neural Comput. App.* **2018,** *30,* 413–435.

Goldberg, D. E. Genetic Algorithms in Search. Optimization, and Machine Learning, 1989.

http://www.scholarpedia.org/article/Ant_colony_optimization

https://heartbeat.fritz.ai/hands-on-with-feature-selection-techniques-an-introduction-1d8dc6d86c16

https://towardsdatascience.com/

https://towardsdatascience.com/the-inspiration-of-an-ant-colony-optimization-f377568ea03f

https://www.analyticsvidhya.com/

https://www.geeksforgeeks.org/introduction-to-ant-colony-optimization/

https://www.neuraldesigner.com/blog/genetic_algorithms_for_feature_selection

https://www.tutorialspoint.com/genetic_algorithms/genetic_algorithms_fundamentals.htm

Joshi, A. S.; Kulkarni, O.; Kakandikar, G. M.; Nandedkar, V. M. Cuckoo Search Optimization-a Review. *Mater. Today Proc.* **2017,** *4* (8), 7262–7269.

Karaboga, D.; Gorkemli, B.; Ozturk, C.; Karaboga, N. A Comprehensive Survey: Artificial Bee Colony (ABC) Algorithm and Applications. *Artif. Intell. Rev.* **2014,** *42* (1), 21–57.

Khan, W. A.; Hamadneh, N. N.; Tilahun, S. L.; Ngnotchouye, J. M. A Review and Comparative Study of Firefly Algorithm and Its Modified Versions. *Optim. Algorithms-Methods App.* **2016,** 281–313.

Lyu, C.; Shi, Y.; Sun, L. A Novel Local Community Detection Method Using Evolutionary Computation. *IEEE Trans. Cybernet.* **2019.**

Mafarja, M.M.; Mirjalili, S. Hybrid Whale Optimization Algorithm with Simulated Annealing for Feature Selection. *Neurocomputing* **2017,** *260,* 302–312.

Meraihi, Y.; Ramdane-Cherif, A.; Acheli, D. Dragonfly Algorithm: A Comprehensive Review and Applications. *Neural Comput App.* **2020,** *32,* 16625–16646.

Mirjalili S. Dragonfly Algorithm: A New Metaheuristic Optimization Technique for Solving Single Objective, Discrete, and Multi-Objective Problems. *Neural Comput. App.* **2016,** *27* (4), 1053–1073.

Rana, N.; Latiff, M. S. A.; Abdulhamid, S. M. Whale Optimization Algorithm: A Systematic Review of Contemporary Applications, Modifications and Developments. *Neural Comput. App.* **2020,** *32,* 16245–16277.

Rostami, M. et al. Review of Swarm Intelligence-Based Feature Selection Methods. *Eng. App. Artif. Intell.* **2021,** *100,* 104210.

Sasikala, S.; Devi, D. R. A Review of Traditional and Swarm Search Based Feature Selection Algorithms for Handling Data Stream Classification. *2017 Third International Conference on Sensing, Signal Processing and Security (ICSSS).* IEEE, 2017; pp 514–520.

Stefanowski, J.; Cuzzocrea, A.; Slezak, D. Processing and Mining Complex Data Streams. *Inf. Sci.* **2014,** *285,* 63–65.

Tereshko, V.; Loengarov, A. Collective Decision Making in Honey-Bee Foraging Dynamics. *Comput. Inf. Syst.* **2005,** *9* (3), 1.

Tommasel, A.; Godoy, D. A Social-Aware Online Short-Text Feature Selection Technique for Social Media. *Inf. Fusion* **2018,** *40,* 1–17.

Wang, J.; Zhao, P.; Hoi, S. C. H.; Jin, R. Online Feature Selection and Its Applications. *IEEE Trans. Knowl. Data Eng.* **2014,** *26* (3), 698–710.

Yang, J. et al. Swarm Intelligence in Data Science Applications, Opportunities and Challenges. In *Advances in Swarm Intelligence*; Tan, Y., Shi, Y., Tuba, M., Eds.; Lecture Notes in Computer Science; Springer: Cham, 2020.

Yang, X. S. Firefly Algorithms for Multimodal Optimization. In *Stochastic Algorithms: Foundations and Applications*; Watanabe, O., Zeugmann, T., Eds., Vol. 5792; Lecture Notes in Computer Science; Springer: Berlin, Heidelberg, 2009.

Yang, X. S.; Deb, S. Cuckoo Search: Recent Advances and Applications. *Neural Comput. App.* **2014,** *24* (1), 169–174.

Yu, K.; Wu, X.; Ding, W; Pei, J. Scalable and Accurate Online Feature Selection for Big Data. *ACM Trans. Knowl. Discov. Data (TKDD)* **2016,** *11* (2), 1–39.

Zhang, Y.; Balochian,S.; Agarwal, P.; Bhatnagar,V.; Housheya,O J. Artificial Intelligence and Its Applications. *Math. Problems Eng.* **2014,** 840491.

CHAPTER 4

Big Data Streams

ABSTRACT

In this chapter, the core methods of big data streams and the prerequisite for parallelization are highlighted. This chapter also discusses about the streaming architecture with an example of Apache Kafka and the benefits of stream processing. Here the fundamentals of the streaming process, its need, and how it performs are discussed briefly. Various real time streaming platforms are listed in this chapter. To comprehend the same, Hadoop architecture is comprehensively explained with the components of parallel processing. The future trends in streaming data is also highlighted that leads to new research challenges.

4.1 INTRODUCTION

Big data streaming is a rapid progression that facilitates a seeker to extract real-time information from any big data. The knowledge obtained from this process thoroughly depends on the rate in which the data are streamed in real time. Big data streaming initiates a rapid data stream that is gathered and stored continuously (Hirzel et al., 2018).

It is a systematic technique that consumes massive streams of real-time data and renders information and patterns that can be utilized for varied purposes. Unless and until you store big data on any specified drive, a constant stream of unstructured data will keep directing to the memory for review. This kind of massive storage becomes possible through a server cluster.

Research Practitioner's Handbook on Big Data Analytics. S. Sasikala, PhD, D. Renuka Devi, & Raghvendra Kumar, PhD (Editor)
© 2023 Apple Academic Press, Inc. Co-published with CRC Press (Taylor & Francis)

In the process of streaming big data, pace counts the most. When the value of results is not subjected to a rapid process, it will eventually decline over time. Timely processing becomes a perplexing task for technical specialists to come up with data stream processing without a server. Now many e-business sectors employ innovative processing tools in big data streaming. This has created a huge demand for technical specialists who are resourceful in providing serverless big data streaming.

Here the fundamentals of the streaming process, its need, and how it performs are discussed briefly. Further, it continues with the way to compose quick and scalable software for the distributed stream processing in <40 lines of code. Since stream processing is a vast topic to summarize in a single chapter, the following sections focus on the data management aspects only.

4.2   STREAM PROCESSING

Let us explore how data processing was performed earlier and then get into the genesis of stream processing. The entire data processed in a database (distributed file system) prior to stream processing is called batch processing (Sun et al., 2015). It is depicted in (Figure 4.1). Here multiple programs compute the generated data. The batch processing software was designed to process finite-size databases. Due to this factor, if some new data emerge, an algorithm starts to crunch the perpetuated data from the recent past, such as in the last one hour or the past day.

Batch processing used to be a very successful tool for many years, and we can still find its implementation on various verticals. But there are some notable disadvantages associated with batch processing. The foremost fact is that the newly emerging data are not subjected to analysis after its instant arrival. This apparently comes up with the following issues:

- High latency fresh outcomes are only measured after a significant interval, but this is unacceptable as the value of data diminishes over time.
- Batch processing method divides data into periods. This restricts session data to provide lucid evaluations. The activities that begins in one stipulated period time but finishes in another time interval are not evaluated to precision.

- Simultaneous processing becomes impossible. Although a block of data accumulation is over completely, it does not get immediately processed instead the batch processing waits for the next block to be gathered completely. Once it is systematically done, then only processing for the preceding block of data is initiated.

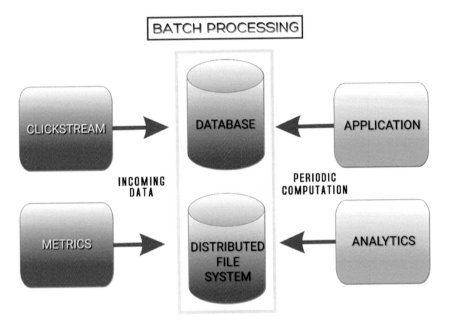

FIGURE 4.1 Batch processing.

Stream retrieval is all about the retrieval of embedded data. A standard stream application consists of many suppliers creating new events and a selection of customers accessing those events. Every detail including financial transfers, user interactions on a platform, or device indicators could be events embedded in the entire framework. Users can aggregate incoming data, send and receive real-time automated updates, or generate new data sources that other users can process. The architecture is shown in Figure 4.2.

There are some highlights from this architecture:

- Low latency: A device may handle and react to new events in real time.

- A natural fit for many systems: Stream processing is a natural fit for applications that run in an infinite stream of events.
- Time investment: Instead of investing time in waiting for the succeeding data to emerge completely, it swiftly kicks starts the processing. So, the data accumulation and processing occur simultaneously. Hence the stream processing device executes calculation as soon as it acquires new knowledge.

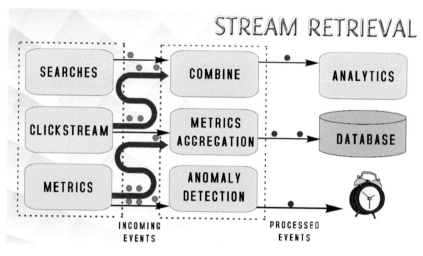

FIGURE 4.2   Stream retrieval.

Unsurprisingly, stream processing was first introduced in real time by finance sectors that need to handle new data instantaneously. Account holder's transactions or Forex rates can be timely updated and processed. They also employ stream processing in identifying fraudulent access, cross-platform recommendations, and customized reports, and so on.

However, this architecture triggers some insightful repercussions in letting manufacturers and consumers to be directly linked with each other. For example, should a manufacturer open a transmission control protocol (TCP) session and submit events directly to every customer? If that is the case, the process becomes impractical and time-consuming.

Despite being an alternative tool, mutual engagement between manufacturers and customers on a common space certainly leads to some serious challenges. What if a manufacturer and customer access the data simultaneously? If the number of customers and producers is exponentially increased, such networks of links are prone to transform into an unruly mess.

For instance, in 2008, LinkedIn engaged several multiple point-to-point pipelines between multiple structures and ended up in some serious troubleshoots. Then specialists began to examine their internal projects to reorganize, which ultimately resulted in a solution called Apache Kafka (Hiraman, 2018). In short, Kafka (Figure 4.3) is a buffer that enables manufacturers to store new streaming events and let the users read them at their own downlink speed, in real time.

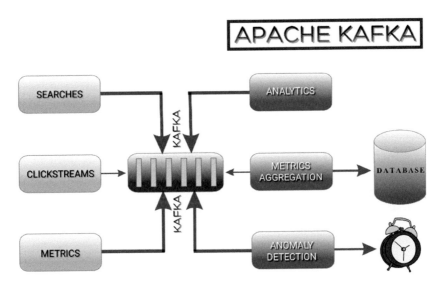

FIGURE 4.3   Apache Kafka.

In offering a data abstraction layer for decoupled event-driven systems, Apache Kafka soon became the cornerstone of modern stream processing applications. It facilitates the possibilities for new buyers and suppliers to be quickly introduced to one another and provides the probabilities to design sophisticated applications.

Apache Kafka is quite popular in this new-age digital era and it successfully shadowed Azure Event Hubs, AWS Kinesis, and other similar architectures. In addition to Apache Kafka, Apache Samza, Apache Flink,

Apache Apex, and Apache Flume are notable popular stream processing frameworks.

4.3 BENEFITS OF STREAM PROCESSING

Stream processing was a "niche" technology used by only a select number of businesses. Industry-wide corporations are now dabbling in streaming analytics, thanks to the rapid growth of software as a service (SaaS), Internet of Things (IoT), and deep learning. It is difficult to find a modern enterprise that does not have an app or platform; as the demand for advanced and real-time analytics grows, the need to incorporate modern data processing has become more commonplace.

Although traditional batch architectures will suffice on smaller scales, stream processing has several advantages that other data platforms cannot match (Zomaya and Sakr, 2017):

- Any information is naturally ordered in this way, allowing it to cope with never-ending event streams. Traditional batch processing tools include stopping the flow of operations, extracting batches of data, and mixing the batches to draw broad conclusions. Although merging and collecting data from multiple sources is difficult in stream processing, it allows us to obtain instant information from large amounts of streaming data.
- To allow real-time data analytics, many businesses use stream processing in real-time or near-real-time. Although real-time analytics is still possible for high-performance storage facilities, the data still lends itself to a stream-processing model.
- Detecting trends in time-series data, such as scanning for changes in website traffic data, necessitates continuous data analysis made more difficult by batch processing, which divides data into batches, meaning that some events are divided into two or three batches.
- Simple data scalability interrupts a batch-processing framework by increasing data volumes, forcing it to include additional services or change in the design. The new stream-processing infrastructure is hyperscalable, capable of a single stream processor working with gigabytes of data per second. Without technology improvements, this helps to comfortably cope with increasing data volumes.

4.4 STREAMING ANALYTICS

Streaming analytics operates by empowering companies to set up data streaming real-time analytics computations from smartphones, social networking, sensors, computers, websites, and much more. Streaming analytics with intuitive standards include quick and appropriate time-sensitive processing coupled with language incorporation. A basic SQL version is used for streaming analytics and eliminates the complexity of stream processing systems.

The massive volumes of knowledge are getting streamed continuously. However, the organizations that operate and interpret such streaming data utilize the knowledge and significantly improve their efficacy.

Real-time streaming and monitoring offer organizations to generate statutory warnings, operational warnings in deactivating account services, fishing real-time fraud, and so on. Streaming analytics is extremely elastic and can manage up to 1 GB/s of high event throughput quickly.

It is secure and, with its built-in recovery features, can help deter data loss. Streaming analytics is a service based on the cloud and is therefore a low-cost solution. The entities are paid in accordance with the units consumed in the streaming process.

In short, streaming analytics provides beneficial market data in motion compared with conventional analytics consuming static data to render reports that are subject to change over time.

4.5 REAL-TIME BIG DATA PROCESSING LIFE CYCLE

The life cycle of real-time big data analysis has five phases (Figure 4.4):

4.5.1 REAL-TIME DATA INGESTION

Heterogeneous data channels consume large data. Often real-time computing paradigms utilize message ingestion storage in the data ingestion phase to serve as a buffer. The data ingestion process tools include Flume and Kafka.

Data storage

Data management process includes the management processes of data systems and streams that have various data structures in real time.

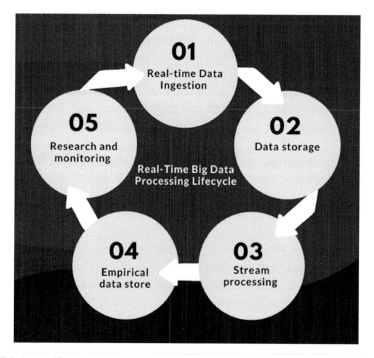

FIGURE 4.4   Life cycle.

Stream processing

The real-time big data stream processing phase is analyzed and organized for the real-time analyst to facilitate decision-making. Different frameworks and paradigms are used at this level, which are based on the design of the real-time application.

Empirical data store

To store and service data in a standardized format, the analytical data store process is always required. The analytic technique is used to find the results from the data store. Resources for the Store Process of Analytical Data: HBase, HIVE.

Research and monitoring

The purpose of analysis and reporting is to include knowledge and perspectives from analysis and reporting for decision-making.

4.5.2 REAL-TIME STREAMING PLATFORMS FOR BIG DATA

Real-time monitoring makes you up-to-date with what is going right now, such as how many people are already reading your recent blog post and how many liked your recent Facebook status. Real time can be functionally efficient but in most of the cases, we cannot get any critical insights (Malek et al., 2017).

Let us presume that you are operating a leading e-commerce portal. Here real-time monitoring could enable us to know whether the new online marketing strategies accelerate or hinder customer activities like billing. The generated reports may prompt to encourage, improve, or change the strategies without investing more time and money.

It also supplies real-time analytics to its own app for additional benefits by suggesting the customer's likes and demands. Let us discuss five large data real-time streaming services further.

Apache Flink

Flink is an open-source streaming application capable of operating pipelines near-real-time, tolerating fault processing, and scalable to millions of instances per second. Flink makes batch- and stream-processing execution.

Apache Spark

Spark is an open-source platform for data processing, which is trending now. Since Spark runs on clusters in-memory and is not related to Hadoop's two-stage MapReduce paradigm, it has lightning-fast efficiency. Spark will operate on Hadoop YARN (Yet Another Negotiator for Resources) as a standalone or on top, where it can read data directly from Hadoop Distributed File System (HDFS). In addition to its processing in-memory, graph processing, and deep learning, streaming may also be done by Spark. It is currently being utilized by organizations such as Yahoo, Intel, Baidu, Trend Micro, and Groupon.

Apache Storm

Storm is a distributed real-time measurement framework that claims to do what Hadoop does for batch processing in streaming. It can be used for analytics, deep learning, ongoing computing, and more in real time. The cool part is that it was built for every programming language to be included. It runs on top of Hadoop YARN and can be used for storing data on HDFS with Flume. The likes of WebMD, Yelp, and Spotify also utilize Storm.

Apache Samza

Samza, based on Apache Kafka and YARN, is a distributed stream-processing system. It offers a quick callback-based API that is close to MapReduce, contains snapshot management, and has fault tolerance.

Amazon Kinesis

Kinesis is a service offered by Amazon for the real-time transmission of streaming cloud data. It is tightly interconnected with other Amazon resources, such as S3, Redshift, and DynamoDB, for full Big Data infrastructure. Kinesis also provides the Kinesis Client Library that helps you to develop apps and use dashboards, warnings, or even dynamic price stream info.

As one of the most promising and key prospects for both academics and business, the task of deriving information from big data has been identified. If the number of applications involving such processing grows, advanced study of large data sources is bound to become a core field of data mining science. One of the key challenges in stream mining is coping with the progression over time of those data sources, that is, with ideas that drift or shift entirely. This section acts as a gentle guide in mining large data streams.

4.6   STREAMING DATA ARCHITECTURE

Streaming data refers to data that are continually generated in large volumes and at a higher rate. A streaming data source is made up of a stream of logs that capture events as they happen, such as a customer clicking on a web page link or a temperature sensor reporting current temperature.

Some examples of common streaming data include:

- Sensors from IoT.
- Logs for servers and security.
- Retirements in real time.
- Data from applications and websites click-stream.

We have sensor devices in all these cases that constantly produce thousands or millions of documents, creating an unstructured or semistructured type of data stream, most usually JSON or XML key-value pairs. A single streaming source can generate thousands of these events per minute. This knowledge is difficult to work within its raw state because it is difficult to query with SQL-based statistical methods due to the lack of schema and structure; therefore, data must be interpreted, evaluated, and structured before serious analysis can be undertaken.

A streaming media infrastructure is a collection of software components that can access and process large volumes of multisource streaming data. Although traditional data systems focus on batch writing and reading data, a streaming data infrastructure dynamically absorbs and stores data as it is generated. This may include things like real-time analytical methods, data mining, and analytics, depending on the use case.

Data streams that tend to contain vast quantities of data (terabytes to petabytes) that are semistructured at best and require significant preprocessing and useful extract–transform–load (ETL) must be considered by streaming architectures.

With a single database or ETL method, stream processing turns out to be a dynamic problem that can be rarely solved. Hence the need to "architect" a solution composed of several building blocks becomes essential.

A streaming data model is an information technology concept that focuses on the processing data in motion and handles ETL batch processing as just another event in a continuous stream of events. An aggregator that collects event streams and batch files from various data sources, a broker that makes data available for consumption, and an analytics engine that analyses data, compare values, and integrates streams together are the three basic components of this type of architecture.

Stream processors are software platforms that enable users to respond to incoming data streams fast. It can send and receive data streams, execute the application, and provide real-time analytics. Since the principle of event sourcing is enabled by a streaming data infrastructure, there is no need for developers to build and manage shared databases.

Instead, all updates to an application's state are saved as a list of event-driven processing triggers that can be restored or queried as needed. When a stream processor receives an event, it reacts in real time or near-real-time by triggering an action, such as retrieving the occurrence for future reference.

In the output of services and products, the growing proliferation of streaming data architectures reflects a shift from a monolithic to a decentralized architecture constructed with microservices. This type of architecture is usually more modular and flexible than a conventional database-based server architecture.

This is because database data processing is colocated to reduce server response times (latency) and improve throughput. Another advantage of using a streaming data architecture is that it considers the moment an event occurs, allowing you to partition and distribute an application's state and processing across several instances.

Developers may use data streaming architectures to produce applications in new ways that provide both bound and unbound data. For example, Alibaba's search technology team updates real-time product details and inventory information using a streaming data platform powered by Apache Flink.

Netflix still utilizes Flink to support its suggestion engines and the architecture is used by ING, the multinational bank headquartered in The Netherlands, to combat identity theft and offer better protection against fraud. Apache Spark, Apache Hurricane, Google Cloud Dataflow, and AWS Kinesis are other platforms capable of handling both stream and batch processing.

4.6.1   THE COMPONENTS OF A STREAMING ARCHITECTURE

Many streaming stacks today provide solutions for problems like stream processing, storage, data collection, and real-time analytics that are based on an open-source and proprietary production line. The modern framework incorporates the bulk of the building blocks and offers a clear way to translate streams into datasets that are ready for study. The components are given in Figure 4.5.

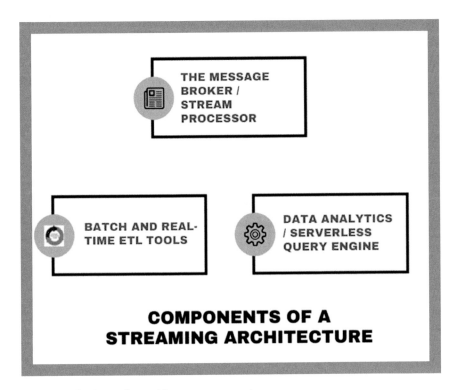

FIGURE 4.5 Streaming architecture components.

4.6.1.1 THE MESSAGE BROKER/STREAM PROCESSOR

This is the segment that takes data from a producer, transforms it into a common message format, and streams it continuously. Other components would then listen to and absorb the signals sent on by the broker (Figure 4.6).

The first component of message brokers, such as RabbitMQ and Apache ActiveMQ, is based on the Message Driven Middleware model. Later on, high-performance messaging networks that are more tailored to a streaming model emerged (often called stream processors). Two common stream processing tools are Apache Kafka and Amazon Kinesis Data Streams.

Streaming brokers support high reliability with durability, have a massive capacity of one gigabyte per second or more of message traffic, and are strictly dependent on streaming with no data transformation or task scheduling support (although Confluent's KSQL can perform basic ETL in real-time while storing data in Kafka).

After studying Apache Kafka data, one can learn about the similarities and differences between Kafka and RabbitMQ, as well as the differences between Apache Kafka and Amazon Kinesis.

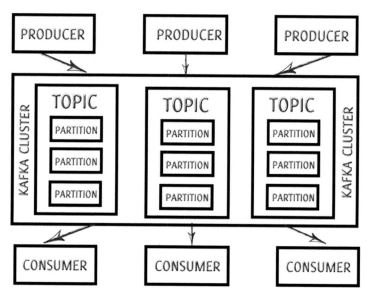

FIGURE 4.6   The message broker.

4.6.1.2   BATCH AND REAL-TIME ETL TOOLS

Data streams from one or more message brokers must be aggregated, transformed, and structured before SQL-based analytics tools can process the data. Mention a few examples, accepting user requests, retrieving events from message queues, applying the response to deliver a reply. Figure 4.7 depicts the database processor. An API request, an incident, a view, an alert, or, in certain cases, a new stream of data may be the result.

Open-source ETL tools for streaming data include Apache Tornado, Spark Streaming, and WSO2 Stream Processor, to name a few. These frameworks, which come in a variety of flavors, can both listen to streams of messages, process the data, and transfer it to disc.

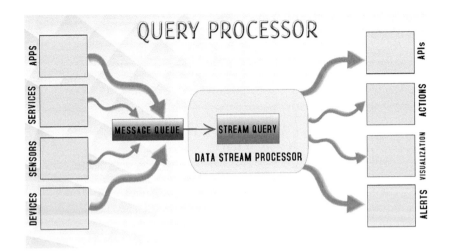

FIGURE 4.7   Query processor.

Few stream processors, such as Spark and WSO2, support SQL syntax for querying and manipulating data, but most operations would require complicated programming in Java or Scala. The ETL data lake was created to provide a self-service solution to converting streaming data using only SQL and a visual interface, rather than the difficulty of Spark's orchestration and control of ETL staff.

4.6.1.3   DATA ANALYTICS/SERVERLESS QUERY ENGINE

The streaming data has to be processed for deriving useful information by the dedicated stream processor. There is a lot of data access streaming across various verticals. Most companies currently store their live event data with the introduction of low-cost computing technologies. Here are a range of streaming information storage options (Figure 4.8), as well as their pros and cons.

A data lake is the most versatile and affordable choice for event data storage, but constructing and managing it is also quite technically involved. The challenges of building a data lake include maintaining best practices for lake storage, as well as the need to ensure proper handling, data partitioning, and historical data backfill. It is simple to dump all the data into object storage; however, constructing an operating data lake can be much more difficult.

DATA STORAGE OPTION	PROS	CONS
DATA BASE / DATA WAREHOUSE	EASY SQL BASED DATA ANALYSIS	HARD TO SCALE AND MANAGE EXPENSIVE
MESSAGE BROKER	EASY TO SET UP NO ADDITIONAL COSTS	DATA RETENTION
DATA LAKE	AGILE LOW COST STORAGE	HIGH LATENCY DIFFICULT TO PERFORM SQL BASED ANALYSIS

FIGURE 4.8   Streaming information storage.

The data lake ETL platform of Upsolver reduces the time-to-value of data lake projects by automating stream ingestion, schema-on-read, and metadata retrieval. This allows data consumers to prepare data with real-time analysis and analytics software more easily.

4.7   MODERN STREAMING ARCHITECTURE

Rather than patching open-source applications together, more organizations are incorporating a full-stack approach in their new streaming data deployments. The modern data paradigm is built on business-centric value chains, where the complexity of traditional technology is abstracted into a single self-service model that converts event sources into analytics-ready data, rather than IT-centric coding processes.

The concept behind Upsolver is to serve as a consolidated data portal that automates the labor-intensive portions of streaming data work: ingestion of texts, batch and streaming ETL, control of storage, and analytics data planning. A modern streaming architecture is shown in Figure 4.9.

The benefits of a digital streaming architecture are various as follows:

- It is possible to eliminate the need for big data innovation programming.
- High availability, performance, and built-in fault tolerance.
- Modern cloud-based services are easy to set up and do not require any upfront investment.
- Adaptability to a variety of use cases and assistance.

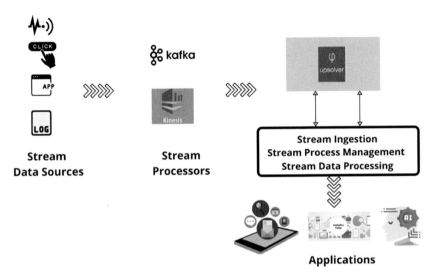

FIGURE 4.9   Modern streaming architecture.

Streaming data analytics can be approached in a variety of ways. Here are some of the most common methods for streaming data analytics.

4.8   THE FUTURE OF STREAMING DATA IN 2019 AND BEYOND

- The design of streaming data is continuous in nature. In 2019 and beyond, three developments we expect would be substantial:
- Streaming data development renders conventional data warehouse systems too inefficient to tackle the increasing adoption of systems that decouple storage and computation. Data lakes are increasingly being used as a low-cost persistence option for storing large volumes of event data, as well as a flexible aggregation point for allowing streaming data access tools outside of the streaming ecosystem.
- Data consumers do not often know the questions they are going to ask in advance, from table modeling to schemaless growth. With as little initial setup as practicable, they want to operate an immersive, iterative process. A burden is lengthy table modeling, schema identification, and retrieval of metadata.

- Data plumbing automation firms are reluctant to devote valuable data engineering time to data plumbing rather than value-adding activities like data cleansing and enrichment. Full-stack solutions are increasingly favored by data departments over personalized home-grown software that reduces time-to-value.

4.9   BIG DATA AND STREAM PROCESSING

The main challenges faced by big data and the solutions for each are listed in Figure 4.10.

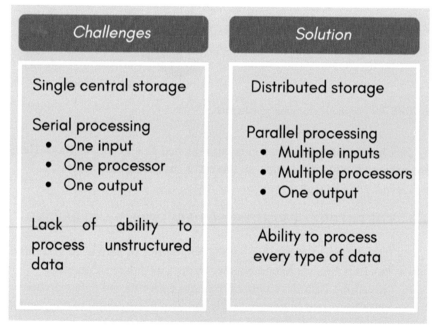

FIGURE 4.10   Big data and stream processing.

4.10   FRAMEWORK FOR PARALLELIZATION ON BIG DATA

In the institutions and user interfaces, parallel processing architectures differ tremendously. To correctly reflect computation, storage, and communication capabilities, along with their interactions, several abstract parallel processing models have been introduced over the years, enabling developers of parallel applications to imagine and take advantage of the available trade-offs without getting bogged down in machine-specific information.

It may separate data-parallel and control-parallel approaches to parallel processing at the coarse stage. Data parallelism means partitioning a broad data collection between several processing nodes before engaging in the process of averaging the partial output, with each running on an allocated chunk of data. It is also the case that the same collection of operations must be done on each data chunk in big data applications, rendering SIMD processing the most powerful option.

However, synchronization with a large number of processing nodes creates overheads and inefficiencies that bite into the gains in pace. Therefore, it may be helpful to guide the processing nodes to asynchronously conduct a single programmed on multiple data chunks, with sparse synchronization, giving birth to the parallel processing model SPMD. MapReduce from Google is the most common cloud-based framework for the parallel processing of SPMD.

Data are spread on various processing nodes in the map stage within individual tasks. A collection of key-value pairs is generated by each map stage. These outputs are then fed to activities that are diminished, where they are merged into a smaller number of key-value pairs that make up the final product. Hadoop on Apache servers is a commonly deployed open-source implementation of MapReduce.

It consists of the MapReduce variant of Hadoop, the distributed file structure of Hadoop, which enables a user to store large data files across several nodes while retaining a centrally placed file picture (stored in one piece), and Hadoop YARN, which handles computational resources and enables activities to be planned on them.

In several respects, the completely open-source Apache Spark is close to MapReduce, with the key differences that it uses a single engine to substitute numerous other required subsystems, rendering the programming effort much more manageable and less vulnerable to error, and an abstraction of data sharing, resilient distributed data sets or RDDs, rendering it substantially more effective for some Multiple nodes work separately on subproblems in control-parallel systems, which are more widely applicable than data parallelism, synchronizing with each other by transmitting input and synchronization messages.

The freedom listed allows for the use of heterogeneous and application-specific resources, without cross-interference or slower resources hindering faster development. Later, more on this. Bulk synchronous parallel, or BSP, processing, is an appealing submodel of control parallelism, which seeks to minimize contact and I/O overheads. BSP equations consist of super steps free of contact and synchronization that are separately performed to conclude before either contact or synchronization takes effect.

Google's Pregel method is a realistic BSP implementation for iterative graph algorithms to be implemented, thus keeping an entire broad graph in memory, distributed over several nodes, to prevent latency of disc access. In event management and stream processing, technologies widely found in social networking and other notification-driven applications, more instances of the control-parallel model are shown.

In large data parallel processing in addition to the two broad forms of parallel processing, reflected in data-parallel and control-parallel structures, there are many other forms of parallelism that may be opposing or complementary methods. This includes parallelism at the stage of instruction, parallelism of subwords, parallelism of data flow, and simultaneous multithreaded parallelism, the latter providing the gain of lower overhead as well as better allocation of capital in operating parallel interdependent tasks.

4.10.1   HETEROGENEITY AND HARDWARE ACCELERATION

Heterogeneous parallel processing requires the use of different processing tools with varying designs and parameters of output. Heterogeneity complicates work distribution and teamwork issues, but it also improves consistency in matching work computational criteria to hardware and software capacities. Users may have access to the resources of a supercomputer in a heterogeneous environment, say, but their usage of such a strong machine is too restricted to warrant a purchase or even a long-term contract of use.

Some nodes can be specialized within a heterogeneous parallel computing framework for extremely efficient execution of such computations utilizing application-specific hardware/software and accelerator augmentation. Examples are use of computing engines for graphic processing units or GPUs, units designed for data storage and data mining, and a range of other advanced tools rendered available across the cloud.

In addition, to set acceleration resources, it is possible to install reconfigurable machines that can be configured to serve as various forms of acceleration mechanisms. Modern FPGAs have enormous computing tools that can be dynamically customized to further reduce bottlenecks in production. Additional possibilities for acceleration are offered by application-specific circuits, whether custom-designed or implemented on FPGAs. Sorting networks, for instance, may be used in a broad variety of ways, including promoting classification or accelerating corresponding search operations.

4.10.2 *INTERCONNECTION NETWORKS*

As computational power, modern parallel computers use commodity processors, frequently multicore, multithreaded, or GPUs. This makes it possible for parallel systems to enable exponential improvements in processing speed and energy consumption accessible instantly. It follows that, to a large degree, the difference between different parallel processing systems is decided by the interconnection networks used to facilitate the sharing of data between computing nodes.

The network linking computing resources, such as Ethernet, may also be a commodity. However, it also pays to build a custom network that fits the connectivity criteria or at least use a custom setup of commodity switches. Numerous interconnection network designs have been proposed and used for multiple device sizes over the years. As computing nodes are attached to the same electronic device, they are linked to an on-chip network, the architecture of which is much more restricted, provided the region, and power limitations.

With several technically superior topologies being unrealistic since they may not be enforced within the physical constraints placed by partitioning, packing, signal delays on long wires, and the like, interconnecting processing nodes within a broad mainframe or supercomputer device pose difficulties in wiring and packing. In terms of topology, interconnecting servers inside a data center pose few constraints, but energy efficiency concerns, as well as stability and serviceability criteria, become dominant given the size.

4.10.3   MAPPING, SCHEDULING, AND VIRTUALIZATION

The mapping of computations to hardware tools is a basic issue in parallel processing. This mapping could preferably be automated and straightforward to alleviate users from having to contend with hardware specifics and adjustments in setup. However, in terms of selective optimizations in adjusting parts of the activities to the capacities of the available computing tools, user-guided mapping contributes to faster running times and more productive usage of resources.

Scheduling is an essential sort of mapping; its origin goes back well before computers arrived on the scene. For parallel processing, scheduling relates to the assigning to processing nodes of computational subtasks when following varying restrictions, in such a manner that an analytical feature is optimized. Task completion period is the easiest target feature, but a few more specific targets, such as reaching deadlines and minimizing energy usage, may also join the picture.

Virtualization makes it possible to use device resources across various users efficiently while isolating and shielding them from each other to maintain privacy and protection. The methods used indicate expansions of the 1960s time-sharing schemes and improvements that fell out of fashion as lightweight and affordable hardware was made available in the form of minicomputers and, finally, microcomputers.

Virtual machine displays separate users from hardware input, resource variability, and shifts in configuration, facilitating greater emphasis on the correctness of the programmed and high-level trade-offs. Reliability is often increased, both since software vulnerabilities are isolated inside virtual machines, avoiding their propagation, and because rescheduling impacted activities on other operating virtual machines will bypass any observed hardware failure.

4.11   HADOOP

Higher volumes and more formats appeared as the years went by and data production improved. Multiple processors were then needed to process information to save time. However, owing to the network overhead that was created, a single storage unit became the bottleneck. This led to a distributed storage device being used by each machine, which made it simpler to access data. This strategy is known as distributed storage parallel processing; different processors execute the processes on separate stores (Widodo et al., 2020).

Hadoop is an Apache open-source application that is used to store and evaluate data processes that are very broad in number. Hadoop is written in Java (online computational processing) and is not Online Scientific Processing (OLAP). It is used by Facebook, Yahoo, Google, Twitter, LinkedIn, and several others for batch/offline production. Moreover, only by attaching nodes to the cluster, it can be scaled up.

4.11.1 FEATURES OF HADOOP

The most common and efficient big data tool is Apache Hadoop, which offers the most secure storage layer in the world. Let us explore different main characteristics of Hadoop in this portion of Hadoop's functionality (Chiang et al., 2021).

Hadoop is Open Source

Hadoop is an open-source project, ensuring that the source code is accessible free of review, alteration, and research costs that allow organizations to change the code according to their specifications.

Hadoop Cluster is Highly Scalable

The Hadoop cluster is scalable, so we can connect any number of nodes (horizontal scalable) or raise the nodes' hardware size (vertical scalable) in order to reach high computational efficiency. This provides the Hadoop system of horizontal as well as vertical scalability.

Hadoop provides Fault Tolerance

The most critical aspect of Hadoop is fault tolerance. A replication function is used by HDFS in Hadoop 2 to have fault tolerance. Depending on the replication factor (by default, it is 3), it generates a copy of each block on the numerous devices. So, if every computer in a cluster goes offline, it is possible to access data from other computers containing the same data as a clone. Hadoop 3 substituted this mechanism of replication with erasure coding. With less room, erasure coding offers the same amount of fault tolerance. The overhead of storage is not more than 50% with Erasure coding.

Hadoop provides High Availability

This Hadoop function ensures high availability, even under unfavorable circumstances, of the results. If one of the Data Nodes fails, the user can still access the data from other Data Nodes that contain a backup of the same data, thanks to Hadoop's fault tolerance feature. The high availability Hadoop cluster, also in a hot standby configuration, consists of two or three operational Name Nodes (active and passive).

The active node is called as the Name Node which is active. The passive node is the standby node that reads the active Name Node edit log modifier and adds it to its own namespace. When an active node crashes, the duty of the active node is taken by the inactive node. Thus, files are available and open to users even though the Name Node goes down.

Hadoop is very Cost-Effective

Since the Hadoop cluster comprises cost-effective commodity hardware nodes, it offers a cost-effective approach for storing and analyzing big data. Hadoop does not require any license to be an open-source product.

Hadoop is Faster in Data Processing

Hadoop manages knowledge in a distributed manner that enables the dissemination of data to be stored on a cluster of nodes. Thus, it provides the Hadoop architecture with lightning-fast processing capabilities.

Hadoop is based on Data Locality concept

Hadoop is popularly recognized for its function of data position, which implies transferring computation logic to the data instead of moving data to the computation logic. This Hadoop function decreases the usage of bandwidth in a system.

Hadoop provides Feasibility

Unlike the normal framework, unstructured data may be processed by Hadoop. It, therefore, provides users with the opportunity to explore data of any format and scale.

Hadoop is Easy to use

As consumers do not have to think about sharing computation, Hadoop is quick to use. The structure itself performs the sorting.

Hadoop ensures Data Reliability

In Hadoop, data is maintained efficiently on cluster machines owing to the redundancy of data in the cluster, amid system failures. The system itself includes a function for the Block Scanner, Volume Scanner, Disk Checker, and Directory Scanner to maintain data durability. If your computer goes down or data is lost, the data is still saved in the cluster reliably and available from the other computer that contains a backup of the data.

4.11.2   MODULES OF HADOOP

1. **HDFS:** Distributed File Structure from Hadoop. Google released its GFS paper and established HDFS based on this. It says that the files in the distributed architecture will be split into blocks and placed in nodes.
2. **YARN:** For task preparation and administration of the cluster, another property negotiator is used.
3. **Map Reduce:** This is a mechanism that enables Java programmed to use a main value pair to calculate data in parallel. Input data is taken by the Map task and transformed into a data collection that can be computed in the Main Value Pair. The Map task output is absorbed by decreasing the task, and then the reducer output produces the desired effect.
4. **Hadoop Common:** To start Hadoop, these Java libraries are used and other Hadoop modules use them.

4.11.3   HADOOP HDFS

In HDFS, data is processed in a distributed way. There are two HDFS components-the node name and the node info. Although there is only one node with a name, there may be many nodes with info. HDFS is specifically built to store large datasets in hardware for commodities. For the complete CPU, a business edition of a cloud costs about $10,000/TB.

This would go up to a million dollars, if you need to purchase 100 of these business edition servers. Hadoop helps us to use computers with commodities as the data nodes. We do not have to waste millions of dollars this way on the data nodes alone. The term node is still a business service.

The HDFS independently stores configuration details and documentation for the file system on dedicated servers. The two critical aspects of the Hadoop HDFS design are the Name Node and the Data Node. User data are stored on servers known as Data Nodes, and information for the file system is stored on servers known as Name Nodes. Based on the replication aspect, HDFS replicates the file content on several storage nodes to guarantee data redundancy (Wadkar et al., 2014).

Using TCP-related protocols, the Name Node and Data Node interact with each other. HDFS must fulfill such prerequisites for the Hadoop architecture to be effective in performance. Both hard drives should have large-throughput speeds. A HDFS file is broken into several blocks, and inside the Hadoop cluster, each is repeated. A block on HDFS is a blob of data with a default size of 64 MB inside the underlying file system. Depending on the specifications, the size of a block may be expanded up to 256 MB.

A distributed file system for Hadoop is the HDFS. It has the design of a master/slave. This design consists of a single Name Node that performs the master function, and a slave function is performed by several Data Nodes.

Name Node and Data Node are also capable of operating on commodity computers. For creating HDFS, the Java language is used. So, the Name Node and Data Node programmed can be conveniently run by any system that supports the Java language.

Name Node

- It operates in the HDFS cluster as a single master node.
- It can become the cause for single point failure, since it is a single node.
- It handles the namespace of the file system by conducting operations such as opening, renaming, and closing data.
- It simplifies the system architecture.

DataNode

- Many DataNodes are part of the HDFS cluster.
- There are several data blocks in each DataNode.

o   To store data, these data blocks are used.
o   Reading and writing requests from the clients of the file system is the duty of DataNode.
o   Upon instruction from NameNode, it performs block formation, deletion, and replication.

Job Tracker

o   The Work Tracker's function is to acknowledge client MapReduce jobs and process the data using NameNode.
o   In response, NameNode supplies the Work Tracker with metadata.

Task Tracker

o   It acts as a Task Tracker slave node.
o   It receives a Work Tracker assignment and code and adds the code to the file. It is also possible to name this method a mapper.

Features of HDFS

•   Delivers distributed storage.
•   Can be applied on hardware for goods.
•   Provides protections for data.
•   Is extremely fault-tolerant because the data from that computer moves to the next computer if one machine goes down.

Master and Slave Nodes

An HDFS cluster is created by master and slave nodes (Figure 4.11). The node name is named the master, and the slaves are named the data nodes.

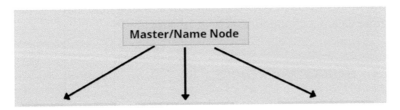

FIGURE 4.11   Master/name node.

The node name is responsible for the functioning of the nodes in the results. It stores metadata, too. The nodes of the data read, compose, process, and repeat the details. They often transmit signals to the called node, known as heartbeats. The state of the data node is indicated by these heartbeats (Figure 4.12).

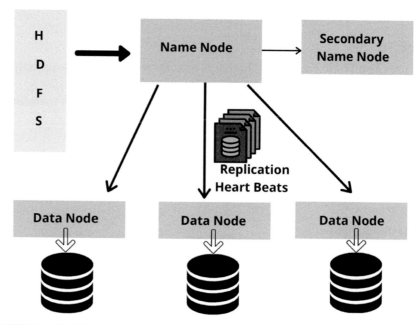

FIGURE 4.12   Name node and data node.

Remember that the Name Node is filled with 30 TB of info. It is spread in the data nodes by the name node and this data is replicated between the data notes. By default, duplication of the data is carried out three times. It is done in this manner, but you can swap it with a new machine that has the same data if a commodity system crashes.

4.11.4   HADOOP MAPREDUCE

The execution device of Hadoop is Hadoop MapReduce. The computation is performed at the slave nodes in the MapReduce approach, and the outcome is submitted to the master node. For the retrieval of the whole data, a data-containing code is used. In relation to the data itself, this coded data is typically quite tiny.

To execute a heavy-duty operation on machines, you only need to submit a few kilobytes worth of code. Based on their names, these key-value pairs are then shuffled and sorted together (Ghazi and Gangodkar, 2015).

The execution of a MapReduce job starts when the client submits a job configuration to the Job Tracker that defines the role map, merge, and decrease, along with the input and output data position. The job tracker identifies the number of splits depending on the input direction upon obtaining the job setup and chooses mission trackers depending on their network proximity to the data sources. The Work Tracker sends a message to the role tracker chosen.

The Map phase analysis starts where the Mission Tracker collects the input data from the splits. With each record parsed by "Input Format" that generates key-value pairs in the memory buffer, the map feature is invoked. By invoking the combine function, the memory buffer is then sorted to separate reducer nodes.

The Role Tracker notifies the Work Tracker upon completion of the chart role. The Work Tracker notifies the selected task trackers when all task trackers are finished to begin the reduction process. The mission tracker reads the files for the area and sorts the key-value pairs for each key. The reduction function, which extracts the aggregated values in the output format, is then invoked.

4.11.5   HADOOP YARN

Hadoop YARN is Hadoop's resource management unit and is available as a Hadoop version two component (Figure 4.13).

- Hadoop YARN serves as an Iso for Hadoop. It is a system of files that is developed on top of HDFS.
- To make sure you do not overwhelm one computer, it is responsible for handling cluster services.
- To ensure that the workers are arranged in the right place, it performs work scheduling

Suppose a client computer needs to run a test or to fetch a data analysis code. This offer for a position goes to the resource manager (Hadoop Yarn), who is responsible for allocating and handling capital. Each of the nodes in the node segment has its own node manager. These node administrators control the nodes and track the node's use of resources. There is a set of physical resources in the tanks, which may be RAM, CPU, or hard drives. The app master demands the container from the node manager if a task request comes through. It goes back to the Resource Manager until the node manager gets the resource.

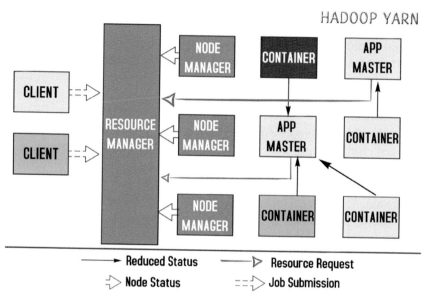

FIGURE 4.13   Hadoop YARN.

4.11.6   *DESIGN OF HADOOP ARCHITECTURE: BEST PRACTICES TO IMPLEMENT*

- To render it cost-effective and scalable to scale up for complex cases in business utilizing, utilizing good-quality commodity servers. Starting with 6 core processors, 96 GB of memory, and 1 0 4 TB of local hard drives is one of the better configurations for the Hadoop architecture. This is just a decent, although not an utter, setup.
- Instead of splitting the two, transfer the computation into near proximity to data for quicker and effective processing of data.
- Using only a Bunch of Disks for redundancy instead of a redundant collection of individual discs, Hadoop scales and works easier for local drives.

- Develop the multitenancy Hadoop architecture by sharing the computing resources with the resources scheduler and sharing HDFS storage.
- Do not edit metadata files as the condition of the Hadoop cluster may be compromised.

Case study

Facebook Hadoop Architecture

With 1.59 billion profiles (about 1/5th of the world's overall population), 30 million FB users change their status at least once a day, 10+ million videos are posted per month, 1+ billion pieces of material are exchanged per week, and more than 1 billion pictures are posted per month, Facebook uses Hadoop to connect with data petabytes.

With more than 4000 computers holding hundreds of millions of gigabytes of files, Facebook operates the world's largest Hadoop Cluster. Facebook's main Hadoop cluster has around 2500 CPU cores and 1 PB of disc space, and Facebook engineers load more than 250 GB of compressed data (more than 2 TB of uncompressed data) into HDFS every day, and there are 100 Hadoop jobs working on these datasets per day.

It checks 135 TB of compressed data daily and introduces 4 TB of compressed data daily. Wondering where all this material is stored? Facebook has a two-level network topology Hadoop/Hive warehouse with 4800 cores, 5.5 PB with up to 12 TB per node stored. With an average of 80 K computing hours, 7500+ Hadoop hive workers are run in output clusters every day. Nonengineers, that is, Facebook analysts use Hadoop via hive and Apache Hadoop jobs are managed by around 200 people every month.

- Facebook's Hadoop/Hive warehouse utilizes a two-level network topology.
- 4 Gbit/s to the rack transfer at the top stage.
- From node to rack transfer, 1 Gbit/s.

Yahoo Hadoop Architecture

Hadoop at Yahoo has 36 separate Hadoop clusters distributed over Apache HBase, Storm, and YARN, with a total of 60,000 servers composed of 100 different hardware configurations installed over decades. Yahoo manages the world's biggest multitenant Hadoop installation with a vast variety of usage cases. Yahoo runs 850,000 volumes of work in a day with Hadoops.

The easiest approach to assess if Hadoop is right for their industry is for companies intending to adopt Hadoop architecture in manufacturing to assess the expense of storing and processing data utilizing Hadoop. Compare the specified expense to the expense of the legacy data storage strategy.

4.11.7 EXPLORE DIFFERENT HADOOP ANALYTICS TOOLS FOR ANALYZING BIG DATA

The Apache Software Foundation created Apache Hadoop, an open-source framework for handling, editing, and analyzing big data. The aim of Hadoop's design was to provide a reliable, cost-effective, and open framework for storing and processing data in a variety of formats and sizes.

In this part, the various Hadoop Analytics resources are discussed. The section outlines the best analytics methods used to process or interpret large data and to gain ideas from it.

Let us now discuss common resources for Hadoop analytics.

Top Hadoop Analytics Tools

1. Apache Spark

It is a well-known open-source big data and deep learning unified analytics engine. Apache Spark was created by the Apache Software Foundation to speed up the collection of big data from Hadoop. It extends Hadoop MapReduce's model to make it easy to use for more calculation modes, such as dynamic queries and stream processing. Over the Hadoop platform, Apache Spark provides batch, real-time, and expert analytics.

Spark performs data analysis in memory for developers and computer analysts. Batch development, joint demands, and broadcasting, among other workloads, have all used it as the default execution engine. Spark has since been introduced on a worldwide scale by companies such as Netflix, Yahoo, and eBay, among many others.

Features of Apache Spark

- **Speed:** Apache Spark can quickly process vast amounts of data thanks to its efficient computing engine. Spark in Hadoop clusters can be 100 times faster in memory and 10 times faster on the disc.

- **Ease of use:** It works with a variety of data stores (OpenStack, HDFS, and Cassandra), making it more robust than Hadoop. Spark offers high-level APIs and facilitates both real-time and batch processing in Java, Scala, Python, and R.
- **Generality:** It contains a library stack, much as Deep Learning MLlib, SQL and Data Frames, GraphX, and Spark Uploading. These libraries can be merged into a single piece of software.
- **Runs Everywhere:** Spark can run on Hadoop, Kubernetes, Apache Mesos, standalone, or in the cloud.

We can explore the additional functionality of Apache Spark.

2. MapReduce

Hadoop is driven by MapReduce. It is a software platform for building applications that process massive datasets on the Hadoop cluster in parallel using hundreds or thousands of nodes. To maximize throughput, Hadoop separates the client's MapReduce job into multiple separate operations that run in parallel. The MapReduce work is split into map and task reduction activities.

Programmers usually write the whole market justification and define the reduction task with lightweight computations like aggregation or summation in the map task. The MapReduce framework functions in two stages: the Map Process and the Reduce Stage. The key-value pair is the input from both phases.

Features of Hadoop MapReduce

- **Scalable:** The MapReduce design is scalable. Until we write a MapReduce programmed model, we will easily expand it to run over a cluster that has hundreds or even thousands of nodes.
- **Fault-tolerant:** It tolerant toward faults and bounces back easily after a loss.

3. Apache Impala

Apache Impala is an open-source framework that improves Apache Hive's performance. It is an Apache Hadoop native analytic database. We can query data stored in HDFS or HBase in real time using Apache Impala. Impala uses the same Apache Hive metadata, ODBC engine, SQL syntax, and user interface, resulting in a common and uniform batch or real-time query structure. We have combined Apache Impala with Apache Hadoop and other leading BI tools to build a cost-effective analytics framework.

Features of Impala

- Security: It is coupled with protection from Hadoop, thereby guaranteeing Kerberos' to be secured.
- Increase the number of Hadoop users: Consumers who use SQL queries or BI apps would have access to more data, thanks to metadata storage from the source through Impala and review.
- Scalability: Linear sizes of Impala, in multitenant environments as well.
- In-memory data analysis: Enables the retrieval of in-memory data, meaning that the data stored on Hadoop DataNodes is easily retrieved and evaluated without any data movement. Thus, it lowers costs because of reduced data movement, modeling, and storage.
- Quicker data access: Compared with other SQL engines, it has a better data access.
- Simple Integration: Impala can be integrated with BI tools such as Tableau, Pentaho, Zoom Info, and others.

We can explore other facets of the Impala as well.

4. Apache Hive

Facebook developed Apache Hive, a Java-based data warehousing framework for the processing and dissemination of massive data. Hive manages vast volumes of data using HQL (Hive Query Language), which is then converted into MapReduce job. It allows developers and analysts to query and interpret big data with SQL-like queries (HQL) without having to write complex MapReduce tasks.

The Beeline shell command-line interface and a JDBC driver are used to communicate with Apache Hive. You may use Hive to analyze or question the massive amounts of data in Hadoop HDFS without having to write complicated MapReduce code.

Apache Hive's Features

- Hive allows you to write client programming in any language, including Python, Java, PHP, Ruby, and C++.

- It usually stores metadata in a relational database management system, which greatly reduces the time needed for textual verification.
- Query efficiency is enhanced by bucketing and Hive Partitioning.
- The Hive is a small, scalable, and scalable ETL tool that allows for faster Online Analytical Processing.
- The User-Defined Utility features are available to assist users with use cases that are not covered by built-in features.

5. Apache Mahout

Apache Mahout is an open-source platform that is widely used in conjunction with the Hadoop architecture to manage massive volumes of data. Mahout is derived from the Hindi word "Mahavat," which literally means "elephant rider". Apache Mahout is made up of algorithms that run on top of the Hadoop method and are called Mahout.

On top of Hadoop, Apache Mahout may be used to implement scalable machine learning algorithms using the MapReduce paradigm. It is a scalable deep learning algorithm library. Previously, the Apache Hadoop framework was used, but now Apache Spark is the predominant platform.

The Hadoop-based implementation of Apache Mahout is not the only option; it can also run algorithms in standalone mode. Basic machine learning algorithms, such as sorting, clustering, suggestion, and mutual filtering, are implemented by Apache Mahout.

Mahout Features

- Since its algorithms are written on top of Hadoop, it fits well in a distributed environment. To scale, it makes use of the Hadoop library in the cloud.
- Mahout provides coders with a ready-to-use platform for performing data mining operations on massive datasets.
- It makes it easy for programmers to evaluate massive datasets.
- Canopy, Mean-Shift, K-means, fuzzy k-means, and other MapReduce approved clustering implementations are used in Apache Mahout.
- Vector and matrix libraries are also used.
- Naive Bayes, Complementary Naive Bayes, and Random Trees are only a few of the classification algorithms introduced by Apache Mahout.

6. Pig

Pig is a different way to help MapReduce work go more smoothly. Pig was developed by Yahoo in order to make MapReduce simpler to write. Pig Latin, a scripting language built for pig architectures and run on the Pig runtime, allows developers to use Pig Latin. Pig Latin is a SQL-like instruction that the programmer converts to MapReduce in the background.

Pig Latin is turned into a YARN MapReduce software for large-scale data processing. This works by loading both the commands and the source of the results. Then we use a number of processes, such as sorting, screening, joining, and so on.

Pig's features include the following:

- Its extensibility where users can write their own functions to process particular intents.
- Pig is ideally suited to addressing complicated use cases that require multiple data assessments and multiple inputs and exports to be processed.
- Pig manages data by making it easy to analyze and archive both structured and unstructured data.
- In Pig, the mission implementation is optimized directly by the work. Therefore, programmers should rely on semantics rather than effects.
- It offers a framework for ETL, loading, and evaluating massive data sets in order to create data flow.

7. HBase

HBase is a distributed open-source NoSQL database that stores sparse data in tables of billions of rows and columns. It is written in Java and based on Google's huge table. HBase, which is built on top of Hadoop, offers support for all data forms. HBase is used, for example, when we have billions of client emails and try to figure out the user name that has replaced the word in their emails, or when we need to look for or recover a limited amount of data from huge data sets. The request must be thoroughly reviewed, and HBase was developed to address those issues.

In HBase, other than the two main components, it contains the following:

- **HBase Manager:** It oversees negotiating load balancing between regional servers. That is not the proper way to treat documents. It keeps track of the failover, manages the Hadoop cluster, and tracks everything.

- **Area Server:** The Zone Server is the worker node that handles interpret, write, delete, and extract client requests. On the HDFS DataNode, the Region Server runs.

HBase features

- Scalable for Storage.
- This facilitates functions that are fault-tolerant.
- Support with the real-time quest for sparse results.
- Easily aid with reading and writing clearly.

8. Apache Storm

Storm is an open-source distributed computational structure written in Clojure and Java that runs in real time. Unbounded data streams (data that expands forever and has no fixed end) can be efficiently processed with Apache Storm. Apache Storm is quick to use and any programming language may be used. For real-time analytics, continuous computing, online machine learning, ETL, and more, we may use Apache Hurricane. Apache Storm is utilized by Yahoo, Alibaba, Groupon, Twitter, Spotify, and many others.

Apache Storm's Features

- It is fault-tolerant and scalable.
- Data processing is assured by Apache Wind.
- It can handle millions of tuples per node per second.
- Setting it up and running it is fast.

9. Tableau

Tableau is a popular data visualization and business solution framework in the business intelligence and analytics area. It is the most effective tool for transforming raw data into an easily readable format that requires no technical knowledge or coding awareness. Tableau helps developers to operate on live databases, invest more time evaluating results, and delivers real-time analytics. Tableau transforms raw data into actionable information and makes decision-making easier. It allows for fast data collection, with visualizations in the form of interactive dashboards and worksheets as a result. It functions in combination with the other big data systems.

Features of Tableau

- Visualizations can be created with Tableau in the form of bar charts, pie charts, histograms, Gantt charts, bullet charts, motion charts, tree maps, boxplots, and several others.
- It is incredibly stable and secure.
- Tableau works with a wide range of data formats, from on-premise libraries, relational databases, spreadsheets, nonrelational databases, big data, data servers, and data in the cloud.
- It allows you to connect with other users in real time and share data in the form of visualizations, dashboards, and sheets, among other things.

10. R PROGRAMMING LANGUAGE

R is an open-source programming language written in C and Fortran. It promotes the use of graphical libraries as well as numerical computing. R can be used for statistical analysis, information analysis, and deep learning. It is platform-agnostic and can be used via a variety of service platforms.

It consists of a comprehensive set of graphical libraries for visually pleasing and elegant visualizations such as plotting, Gplotting, and more. The R language is often used to produce mathematical applications and data processing by statisticians and data miners. The vastness of its kit environment is R's greatest benefit. R promotes the output of multiple mathematical activities and allows to generate findings in text and graphical format for data processing.

R's characteristics

- R has a vast range of packages to select from. It has CRAN, a package repository of over 10,000 packages.
- R offers cross-platform features. It can operate on every operating system.
- R is a word that has been translated into another language. It is not necessary to use a compiler to compile the code. The R script, therefore, flies in a short amount of time.
- Both formal and unstructured information can be accommodated by R.
- R offers unrivalled graphics and device benefits as well as other R attributes.

11. Talend

Talend is an open-source framework for simplifying and automating the integration of big data. It offers numerous data integration tools and facilities, large data, data protection, data quality, and cloud computing. It allows organizations to make choices in real time and to become more data-driven. Talend brings together a number of connectors under one roof, helping us to customize the solution as desired.

Talend Large Data, Talend Data Consistency, Talend Data Aggregation, Talend Data Planning, Talend Server, and other commercial products are available. Talend is used for companies including Groupon, Lenovo, and others.

Features of Talend:

- ETL and ELT are streamlined for large data by Talend.
- It achieves Spark's pace and scope.
- It manages multisource results.

12. Lumify

Lumify is an open-source framework for the fusion, study, and simulation of big data that facilitates the growth of actionable knowledge. Lumify's analytical options, which include full-text faceted search, 2D and 3D graph visualizations, immersive geospatial views, interactive histograms, and real-time shared collaborative workspaces, empower users to discover complex connections and evaluate interactions in their performance.

We can get multiple options on the graph using Lumify to analyze the relationships between individuals. Lumify includes the fundamentals of visual, video, and textual content ingest processing and GUI.

Features of Lumify

- The infrastructure of Lumify requires new analytical instruments to be attached that will function in the background to track adjustments and support analysts.
- It is stable and scalable.
- Lumify offers support for a cloud-based environment.

- Lumify makes it simple to integrate any clear layer-compatible mapping service, such as Google Maps or ESRI, into geospatial analysis.

13. KNIME

Konstanz Details Miner (KNIME) is an open-source, customizable framework for processing big data, data mining, market analysis, text mining, analysis, and business intelligence for data analytics. KNIME allows users to evaluate, manipulate and model data through visual programming. KNIME is a successful SAS substitute.

It includes computational and mathematical features, algorithms for machine learning, sophisticated predictive algorithms, and more. KNIME is used by several companies, including Comcast, Johnson & Johnson, Canadian Tire, among others.

KNIME's features

- KNIME allows for fast operations with ETLs, and it can be easily integrated with other languages and technologies.
- It has over 2000 modules, a wide range of advanced software, and complex algorithms.
- KNIME is easy to install and has no performance issues.

14. Apache Drill

It is a low-latency distributed query engine inspired by Google Dremel. Without having to fix a schema, Apache Drill allows developers to use MapReduce or ETL to explore, simulate, and query large datasets. It is made to scale and interrogate petabytes of data across thousands of nodes.

Just by defining the path to a Hadoop directory, NoSQL folder, or Amazon S3 bucket in the SQL query can we query data with Apache Drill. There is no need for developers to code or build games with Apache Drill. Users can view data in any format from any data source using standard SQL queries.

Apache Drill's Advantages

- Using the high performance, easy-to-use Java API, it is now simpler for developers to reuse their most recent Hive implementations.
- Using the high performance, simple-to-use Java API, it is now easier to create UDF.

- Drill users are not required to construct or handle tables in the metadata to perform a query on results because Apache Drill has a specialized memory management system that eliminates garbage collections and optimizes memory allocation and usage.

15. Pentaho

Pentaho is a platform to transform large data into big ideas with a mantra. It is data automation, orchestration, and a forum for enterprise analytics that offers help ranging from collection, planning, integration, review, estimation, to immersive simulation of big data.

Pentaho develops software for real-time data analysis to improve digital insights. It enables businesses to evaluate and gain information from big data, which encourages businesses to build a productive consumer experience and operate their companies more efficiently and cost-effectively.

Features of Pentaho

- Pentaho can be used for analytics of large data, embedded analytics, analytics of clouds.
- Pentaho advocates OLAP.
- One may use Pentaho for Statistical Research..
- It has a user-friendly interface.
- Pentaho offers solutions for a large variety of big data outlets.
- It helps companies by detailed reporting and dashboards to evaluate, incorporate, and display results.

KEYWORDS

- **big data streams**
- **stream processing**
- **streaming analytics**
- **data ingestion**
- **streaming platforms**
- **streaming data architecture**

REFERENCES

Chiang, D. L.; Wang, S. K.; Wang, Y. Y.; Lin, Y. N.; Hsieh, T. Y.; Yang, C. Y.; Ho, H. Modeling and Analysis of Hadoop MapReduce Systems for Big Data Using Petri Nets. *Appl. Artif. Intell.* **2021,** *35* (1) 80–104.

Ghazi, M. R.; Gangodkar, D. Hadoop, MapReduce and HDFS: A Developers Perspective. *Procedia Comput. Sci.* **2015,** *48,* 45–50.

Hiraman, B. R. A Study of Apache Kafka in Big Data Stream Processing. In *2018 International Conference on Information, Communication, Engineering and Technology (ICICET) IEEE,* 2018; pp 1–3.

Hirzel, M. et al. Stream Processing Languages in the Big Data Era. *ACM Sigmod Record* **2018,** *47* (2), 29–40.

http://spark.apache.org/

https://drill.apache.org/

https://github.com/lumifyio/lumify

https://hbase.apache.org/

https://help.pentaho.com/\

https://hive.apache.org/

https://impala.apache.org/

https://mahout.apache.org/

https://pig.apache.org/

https://storm.apache.org/

https://towardsdatascience.com/introduction-to-stream-processing-5a6db310f1b4

https://www.bmc.com/blogs/hadoop-apache-yarn/

https://www.confluent.io/blog/event-streaming-platform-1/

https://www.dezyre.com/article/hadoop-architecture-explained-what-it-is-and-why-it-matters/317/

https://www.eckerson.com/articles/planning-your-data-architecture-for-streaming-analytics

https://www.geeksforgeeks.org/hadoop-history-or-evolution/

https://www.infoq.com/articles/how-to-choose-stream-processor/

https://www.knime.com/

https://www.r-project.org/

https://www.tableau.com/

https://www.talend.com/

https://www.upsolver.com/blog

https://www.upsolver.com/blog/streaming-data-architecture-key-components\

https://www.wowza.com/blog/future-of-streaming-2020-and-beyond

Malek, Y. N.; Kharbouch, A.; El Khoukhi, H.; Bakhouya, M.; De Florio, V.; El Ouadghiri; Blondia, C. On the Use of IoT and Big Data Technologies for Real-Time Monitoring and Data Processing. *Procedia Comput. Sci.* **2017,** *113,* 429–434.

Sun, D.; Zhang, G.; Zheng, W.; Li, K. Key Technologies for Big Data Stream Computing. In *Big Data Algorithms, Analytics and Applications*; Li, K., Jiang, H., Yang, L. T., Guzzocrea, A., Eds.; Chapman and Hall/CRC: New York, 2015; pp 93–214. ISBN 978-1-4822-4055-9.

Wadkar, S.; Siddalingaiah, M.; Venner, J. *Pro Apache Hadoop*; Apress, 2014.

Widodo, R. N. S.; Abe, H.; Kato, K. HDRF: Hadoop Data Reduction Framework for Hadoop Distributed File System. In *Proceedings of the 11th ACM SIGOPS Asia-Pacific Workshop on Systems*, 2020; pp 122–129.

Zomaya, A. Y.; Sakr, S., Eds. *Handbook of Big Data Technologies*, 2017.

CHAPTER 5

Big Data Classification

ABSTRACT

In this chapter, the big data classification techniques and various learning methodologies are explained with examples. To extend the same, deep learning algorithms, and architectures are also briefed. The machine learning is auto extraction of information with the continuous learning experience without explicit programming. The machine learning types are supervised, unsupervised, semi-supervised, and reinforcement. Various algorithms under each learning types are discussed with examples. Incremental learning and deep learning methods are particularly used for big data streams. To continue with the same, the deep learning algorithms are also covered in this chapter.

5.1 CLASSIFICATION OF BIG DATA AND ITS CHALLENGES

Upsurge in the development of new hardware and software technologies in the current environment has led data to be streamed everywhere and anywhere. But the task to establish storage, process, and project these data is tedious. The major challenge in data mining is the classification of the big datasets.

The conventional classification methods that run in a streaming environment use high memory and has longer execution time. In this chapter, the challenges of classifying big data using machine learning (ML) techniques are outlined with an emphasis on modeling and algorithms, batch learning, and online supervised learning (regression and classification) and unsupervised learning.

Research Practitioner's Handbook on Big Data Analytics. S. Sasikala, PhD, D. Renuka Devi, & Raghvendra Kumar, PhD (Editor)
© 2023 Apple Academic Press, Inc. Co-published with CRC Press (Taylor & Francis)

Traditionally, ML techniques that were developed used labeled datasets to train and validate them for extracting useful information. The following points are considered as important to solve big data classification problems.

1. An ML technique trained on a certain labeled dataset or data domain may not be suitable for another dataset or data domain, that is, the classification may not be robust or appropriate over diverse datasets or data domains.

2. In general, an ML technique is trained with certain types of classes. Therefore, when the same is applied on a dynamically increasing dataset with more types of classes, it leads to inaccurate classification results.

3. An ML technique is developed based on a single learning task. So, they may not be suitable for multiple learning tasks and knowledge transfer requirements of today's big data analytics.

4. Due to quick data arrival, the demand in processing big data of streaming nature is difficult and inflexible.

It is understood that working on fast streaming continuous input data is vital. But there is a need to work on such data analysis as they are useful in real-time application. So, it is important to adapt deep learning (DL) and also look for algorithms that can handle large input data volumes. In this section, the applications of DL with streaming data, including incremental learning and classification are discussed.

Big data analytics and DL are vital aspects of data science (Jan et al., 2019). Big data has influence in many public and private administrations where huge domain-specific information is collected and therefore used in extracting needful information on security, fraudulent activity, marketing, and health care. For example, Google and Microsoft use the same for their business analysis and decision-making.

DL algorithms use a structural hierarchical way of extracting the data representations as high end and complicated constructs. At a particular level, these abstractions are learnt based on quite simpler abstractions formulated at the level of each hierarchical stage. DL is a promising approach where raw, unsupervised, unlabeled, uncategorized, and massive amounts of data are analyzed and learnt.

With such significant advantage, let us dive into how DL can be exploited to process big data and infer the insights such as semantic indexing, tagging of data, quicker retrieval of data, simplifying the complexity of data extraction tasks are detailed out in this section. Precisely, the challenges of big data analytics scalability, streaming data issues, and dimensionality are also elaborated from the research perspectives.

5.2 MACHINE LEARNING

Machine learning or ML, an application of artificial intelligence (AI), enables autoextraction of information with the continuous learning experience without explicit programming. In this context, learning means identifying and understanding input data and making intelligent decisions based on the provided data and its experience. It can be said as the user-generated system that has the capability of accessing the data and learn from them by themselves.

5.2.1 A WALKTHROUGH

Figures 5.1 and 5.2 depict the cycle of ML process. The process starts with data acquisition followed by leaning mechanism which is continued for some iterations to gain experience, and from the learned experiences, predictions are made at the end of this process of learning (Kelleher et al., 2020).

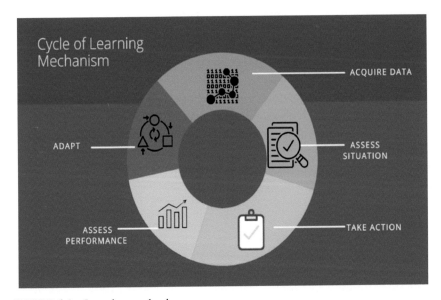

FIGURE 5.1 Learning mechanism.

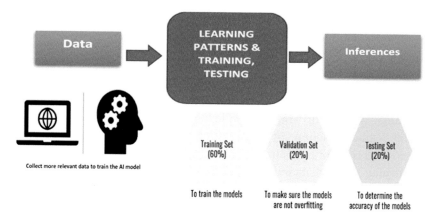

FIGURE 5.2   Machine learning cycle.

Generally, 60% of the data are used to train model to learn the patterns. After training 20% of the datasets are used for validation, then remaining 20% is used for testings the model for accuracy. The ML is trained for considerable amount of time so that it can make prediction with less error. The model is tested with metrics designed for particular problem.

The applications of ML from automated cars to "Alexa" are making day-to-day life interesting. The idea behind the concept has been evolving for many years now and is not very new. Figure 5.3 outlines the milestones.

5.2.1.1   *MACHINE LEARNING IN THE PRESENT SCENARIO*

ML is advancing tremendously and is present around us in self-driving cars, Amazon Alexa, chatbots, recommender systems, and so on. It includes supervised, unsupervised, and reinforcement learning with clustering, classification, decision tree, SVM algorithms, and so on. Modern ML models can make predictions on various applications for weather, health care, and so on.

5.2.2   *TYPES OF ML*

ML algorithms can be categorized into either supervised or unsupervised. Supervised algorithms apply learning from the past to the new data by using class labels to predict future actions. The learning algorithm predicts by analyzing the known training dataset.

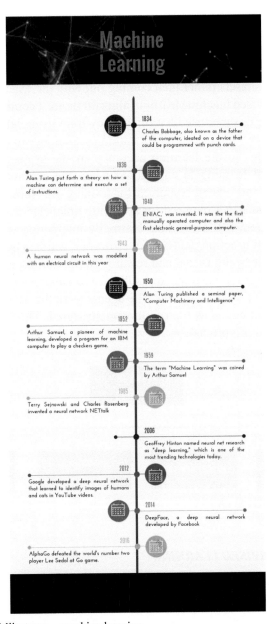

FIGURE 5.3   Milestones—machine learning.

Once the system is adequately trained, it will be able to provide results for any new input with the help of the learning algorithm. The algorithm also helps to modify the model by comparing the delivered output with the intended output and finds errors accordingly.

Unsupervised ML algorithms, in contrast, use information with unknown labels. They study on how the function is inferred to extract the unknown from unlabeled data. It is concerned with the exploration and inferences of datasets rather than coming out with the exact output.

Semisupervised machine learning algorithms are a combination of both supervised and unsupervised learning. They train using labeled and unlabeled data where a major part of the training set belongs to the latter. These algorithms improve the learning accuracy of the systems significantly. Semisupervised learning is considered when the acquired labeled data require expert and related resources for training and subsequent learning which otherwise is not required for acquiring unlabeled data generally.

Reinforcement ML algorithms learns by interacting with an external entity constructing actions and notices errors or rewards. The characteristics of this type of learning is trial and reward. To maximize performance, this approach enables the machine and software to autodetect the appropriate behavior within a specific setting. The agents learn the best action through a reward system called as the reinforcement signal. The overview of the types is given in Figure 5.4.

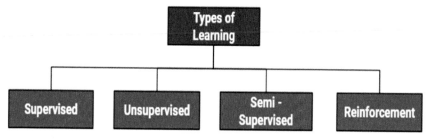

FIGURE 5.4   Types of ML.

5.2.3   SUPERVISED LEARNING

In supervised learning, sample labeled data is fed to the learning system for training which in turn predicts the output. It is synonymous to a classroom supervision where the student learns something new from the teacher who is already aware of the same. A model is created by the system by learning every labeled data in the datasets. After training the model, it is tested for prediction accuracy using sample data.

The aim of this type of learning is to correlate with input to output data. Consider, there are input variables X, Y is the output class variable and procedure is used to correctly relate the output for the input $Y = f(X)$.

The objective is to approximate the mapping function in such a way that, for the newly given input data (X), the outcome is predicted as (Y). This process exits when the learning algorithm reaches the enhanced accuracy level and ready for further predictions. The divisions of supervised learning problems are presented in Figure 5.5.

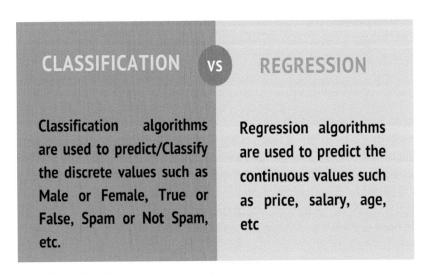

FIGURE 5.5   Classification versus regression.

Different algorithms and computation methods are used in supervised ML such as,

- Regression
- Logistic Regression
- Classification
- Naive Bayes Classifiers
- k-NN (k-nearest neighbors)
- Decision Trees
- Support Vector Machine

The following sections explain the most used supervised learning methods.

5.2.3.1 NEURAL NETWORKS

Neural networks were largely leveraged for DL algorithms (Aggarwal, 2018). The data is trained through the layers of node. Each node consists of inputs, weights, a bias (or threshold), and an output. If the output value exceeds the threshold, it activates and fires the node which in turn, passes the input into the next layer in the architecture. This mapping is learnt by the neural networks through supervised learning, where the adjusting is based on the loss function through gradient descent. The model is accurate when the cost function reaches zero value.

5.2.3.2 NAIVE BAYES

Naive Bayes is based on classification and built on the principle of Bayes Theorem (Xu, 2018) (Figure 5.6). That is, each predictor has the same effect on the result and the incidence of one feature is not correlated with the incidence of other in the probability of a given outcome.

There are three types of Naive Bayes classifiers: Multinomial Naive Bayes, Bernoulli Naive Bayes, and Gaussian Naive Bayes. This classifier has been used in classifying the text and providing recommendations.

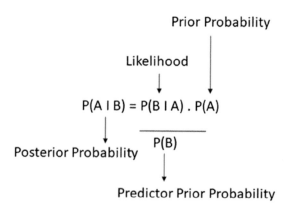

FIGURE 5.6   Naive Bayes theorem.

5.2.3.3   LINEAR REGRESSION

Linear regression (LR) relates the dependent and independent variable and establishes the correlation strength between them. It is usually leveraged to make future outcome predictions. In the case of simple LR, there is linear relationship between single dependent and independent variable.

It becomes a multiple linear regression, as the number of independent variables increases. Every type of linear regression uses the least squares method to fit a best line (Figure 5.7). However, the projected graph contains the straight line of fit, unlike other regression models.

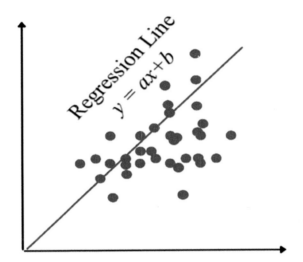

FIGURE 5.7   Linear regression.

5.2.3.4   LOGISTIC REGRESSION

LR is considered when dependent variables are continuous whereas logistical regression is taken when the dependent variable can be categorized, that is, when they have binary outcome of either "true" and "false" or "yes" and "no." Though both regression models seek to identify relationships between input data, logistic regression is primarily used to solve binary classification problems.

The logistic function is represented in Figure 5.8. The curve raises from 0.0 to 1.0 is called as the Sigmoid curve. Thus, the curve is S shaped called as Sigmoid S-shaped curve. When $z \to \infty$, the output range of the curve is 1, and is 0 when $z \to -\infty$. The range of Sigmoid/logistic function lies between zero and one (probability of being in a class or not).

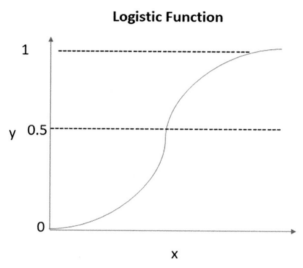

FIGURE 5.8   Logistic function.

5.2.3.5   SUPPORT VECTOR MACHINE

Support vector machine (SVM) developed by Vladimir Vapnik (Boser et al., 1992), is a popular supervised learning model used for both data classification and regression. It is typically used on classification problems wherein it constructs a hyperplane that is considered as the decision boundary.

From Figure 5.9, the hyperplane is drawn between two classes A and B at the maximum distance, thus dividing both the classes on either side of the plane.

5.2.3.6   K-NEAREST NEIGHBOR

K-nearest neighbor (KNN) algorithm (Cunningham and Delany, 2020) is a nonparametric algorithm that categorizes data points connected with closeness and correlation with the other available data. It is based on the hypothesis that similar data are found nearer.

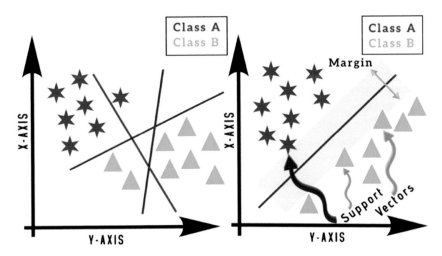

FIGURE 5.9   Support vector machine.

The nearest neighboring points are found by the Euclidian distance (Figure 5.10). KNN is the most preferred by data scientists because of its ease of use and low calculation time but when it comes to classification with a growing dataset, processing takes time. KNN is typically used for image classification and recommendation systems.

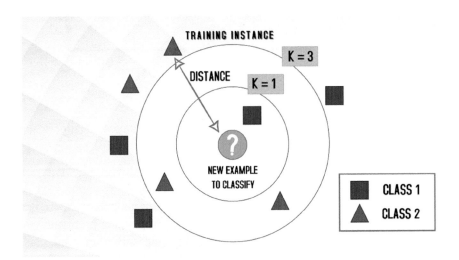

FIGURE 5.10   K-nearest neighbor.

5.2.3.7 RANDOM FOREST

Random forest, a yet another adaptable supervised ML algorithm is used for classification and regression. In this context, "forest" comprises unrelated decision trees, where they are put together for variance reduction and accurate data prediction. The random forest algorithm (Figure 5.11) combines the result of several decision trees by means of voting rule and consolidates the final outcome.

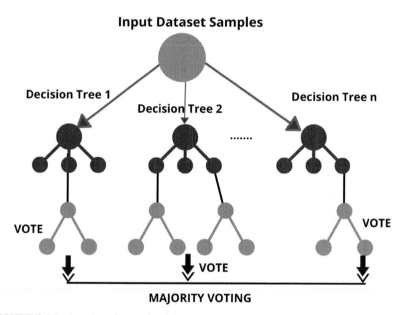

FIGURE 5.11   Random forest algorithm.

The following are the advantages and disadvantages of supervised learning:

> ➢ With supervised learning, data can be collected with outputs based on previous experiences.
> ➢ It helps to enhance performance criteria with the support of experience and gives solutions to various real-world computation problems.
> ➢ On the flip side, training requires high computation time and classification of big data remains challenging.

The following section explains unsupervised learning.

5.2.4 *UNSUPERVISED LEARNING*

Unsupervised learning is ML without any supervision. It is synonymous to self-learning when students try to learn by themselves about the new and unknown data in hand without a teacher. The students are able to infer from the data based on their own thinking and do not require any affirmation. Here, the machine is trained with the set of unlabeled, unclassified, or uncategorized data using the algorithm that acts on the data without any supervision.

The idea of this learning is to identify the new features or group of similarly patterned objects; thus modeling and restructuring the input data. Doing so, the machine attempts to learn useful inferences from the huge volume of data rather and is not concerned on bringing the exact output. It depends on the algorithm to know and give an interesting structure to the data.

Unsupervised learning can be classified into clustering and association, which is depicted in Figure 5.12.

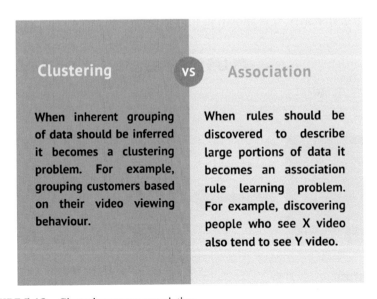

Clustering VS **Association**

| When inherent grouping of data should be inferred it becomes a clustering problem. For example, grouping customers based on their video viewing behaviour. | When rules should be discovered to describe large portions of data it becomes an association rule learning problem. For example, discovering people who see X video also tend to see Y video. |

FIGURE 5.12 Clustering versus association.

The following sections discuss about the types of unsupervised learning.

Clustering

> ➤ Exclusive (partitioning)
> ➤ Agglomerative
> ➤ Overlapping
> ➤ Probabilistic

Clustering Types

> ➤ Hierarchical clustering
> ➤ k-means clustering
> ➤ Principal component analysis
> ➤ Singular value decomposition
> ➤ Independent component analysis
> ➤ Clustering

Clustering, an important concept in unsupervised learning, works on discovering patterns or structures from uncategorized data. They process data and attempt to find natural clusters or groups, if exist. The number of clusters to be identified can be modified and also allows to adjust the group's granularity. It is shown in Figure 5.13.

CLUSTERING

Input   **Clusters Same Type**

FIGURE 5.13   Clustering.

The different types of clustering (Saxena et al., 2017) are as follows:

Exclusive (partitioning)

In this type of clustering, data grouping is done in such a way that data can be part of only one cluster. For example, k-means clustering.

Agglomerative

In agglomerative clustering, each data point is considered as a cluster. The adjacent clusters are iteratively combined in order to reduce the number of clusters. Hierarchical clustering uses this technique.

Overlapping

Here, clustering is done using fuzzy sets. A data point can participate in more than one cluster with individual degree of membership to whose membership value, the data will be associated. For example, fuzzy c-means clustering.

Probabilistic

Clustering is done using probability distribution. Consider the following keywords—"boy's bag," "girl's bag," "boy's bottle," "girl's bottle." Using this technique, the data can be clustered into two categories, "boy," "girl," or "bag," and "bottle."

The following section explains about the clustering types. They are as listed below.

- Hierarchical clustering
- k-means clustering
- Agglomerative clustering
- Dendrogram
- Principal component analysis

Hierarchical Clustering

In this approach, clusters are formed in a hierarchical structure. Clustering starts with the data that is part of their own cluster. The adjacently placed clusters will be in the same cluster and the algorithm stops when only one cluster is remaining.

k-means clustering

This algorithm starts with a number of clusters (k) such that the data points are gathered into "k" groups. The objective to find the highest k value that denotes smaller groups with more granularity and lower "k" value denotes larger groups with less granularity.

The algorithm outputs a group of labels. It assigns every data point to one of the desired "k" groups where each group has a centroid, the heart of the cluster. These centroids capture the data points in proximity and puts them into their cluster. k-means clustering can be further downed to two subgroups namely agglomerative clustering and dendrogram.

Agglomerative Clustering

This clustering takes each data as a single cluster and does not require "k," the number of clusters. It starts with a fixed number of clusters and allots all data points to the exact number of clusters. By merging based on some distance measure, the number of clusters are merged. At the end, there is only one big cluster with all the objects.

Dendrogram

Here, every level will denote a possible cluster where the dendrogram height symbolizes the similarity level between the clusters. So, finding groups from dendrogram is mostly subjective and not natural.

Principal Components Analysis

This is used when there is a need for higher dimensional space. A basis is selected from the space with the 200 most important scores for the same. This is identified as the principal component. A new subset with a new space smaller in size to the initial space is selected which maintains the complexity of the data as much as possible.

Association

In order to bind the data in large databases, association rules are required. This technique discovers remarkable relationships between data in large databases. For example, inferences like people that buy a new painting canvas are most likely to buy related stationery can be learnt.

Other examples are grouping of COVID 19 patients based on their CO-RADS score, grouping of customers based on their previous browsing and shopping cart entries, grouping of songs based on the streaming numbers and so on.

Applications of Unsupervised ML

- In clustering, based on similarities the dataset is split into groups and clustered automatically.
- Anomaly detection is used in uncovering fraudulent transactions as they can discover unfamiliar data points in the datapoints.
- Association mining helps to identify the sets of items which are often together in the dataset.
- Latent variable models are mostly used for data preprocessing methods like reducing the features and disintegrating the dataset into multiple components and so on.

Since, unsupervised method can learn without class label, it can solve more complex tasks compared to supervised learning. Also, getting unlabeled data is comparatively easy than getting labeled data.

On the flipside, unsupervised learning is comparatively complex than supervised learning since the output is not definite. Since the input data is unlabeled and do not know the result beforehand, the output of an unsupervised learning algorithm might not be fully accurate.

5.2.5 SEMISUPERVISED ML

Semisupervised ML involves large volume of input data (X) with partly labeled data (Y). It can be said as a mix of both supervised and unsupervised learning. Consider a map archive where known locations are labeled, and unknown locations are left unlabeled. Many such real-world problems exist where it can be quite expensive and time-consuming as they require expert opinion on labeling which is not required for unlabeled data.

Unsupervised learning techniques are applied to infer and learn the structure of input data; supervised learning can also be applied to get predictions on unlabeled data. This information is then fed to the supervised learning algorithm to come up with a model for predicting on new unknown data.

5.2.6 REINFORCEMENT LEARNING

Reinforcement learning learns by the feedback system that provides reward or penalty based on the correct and incorrect actions. The agent progresses its performance by automatically learning from these feedbacks; more the reward points better the performance. It explores the environment by interaction and the goal is to get the most reward points. One example of reinforcement learning is a robotic arm, which automatically learns its movements. The key components are depicted in Figure 5.14.

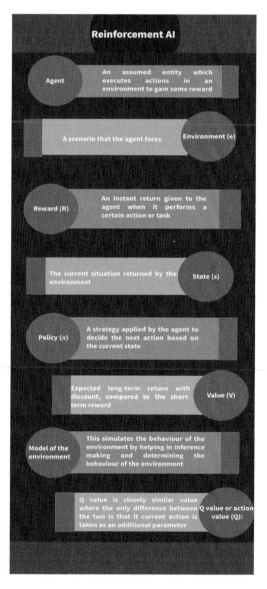

FIGURE 5.14 Reinforcement learning components.

Reinforcement Learning Working Mechanism

For case study, we will take training pet with new tricks. As it cannot understand any human language, the mode of communication and the strategy used should be different. The pet is placed in a situation and look for different responses it gives. If one of the responses is the expected response, it is given a treat. Once again, when the pet is put up in the same situation, it gets excited to give the same response as before as the pet knows it will be rewarded again with a treat. This way, the pet learns from the positive experiences (being rewarded) to set a correct action and avoid particular actions from the negative experiences (not being rewarded). The scenario is illustrated in Figure 5.15.

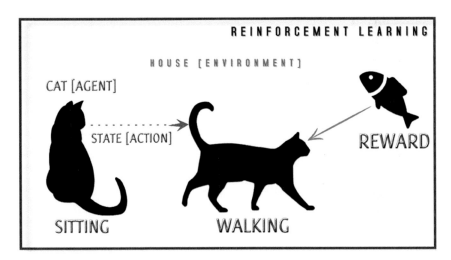

FIGURE 5.15 Reinforcement learning.

Here, the pet is the agent exposed to the environment and house. The current state is the pet sitting and a specific word can be said for the pet to go to a different state, for example, walk. The agent does an action where it transits from one state (sitting) to another (walking).

The reaction of the agent becomes the action that it does, and the policy shows the method of selecting the action that yields an enhanced result. Once the transition of state is completed by the agent, it is rewarded by a treat (right action) or given a penalty (wrong action).

Reinforcement Learning Algorithms

The different way of implementing Reinforcement Learning algorithms is explained below.

Value Based

The main focus in value-based type is explore the state and action value. The maximum valued action is taken.

Policy Based

In this method, a policy must be devised to gain the maximum reward for every action taken in a state.

Model Based

This approach constructs a model for the agent to learn. The model will guide the agent in prediction. The following are the characteristics of reinforcement learning.

Characteristics of Reinforcement Learning

The following are the characteristics of reinforcement learning:

- It does sequential decision-making.
- Time is vital in this type of learning.
- Feedback is not instant; they are constantly delayed.
- The actions of the agent determine the next data it will get.

Types of Reinforcement Learning

The types of reinforcement learning methods are discussed below.

Positive

It is defined as an event which happens due to certain behavior. The strength and frequency of the behavior is increased which has a positive impact on the action taken by the agent. Such reinforcement helps in maximizing the performance and sustaining change for an extended period. But, using reinforcement way too much may lead to overoptimization of a state which affects the results.

Negative

It is defined as strengthening of behavior that happens due to a negative condition that should have been stopped or avoided. This helps in defining the minimum level of performance. The minimum performance behavior is considered as a drawback.

Reinforcement Learning Models

The key types of learning models are given below,

- Markov Decision Process
- Q learning

Markov Decision Process

The different components of this model are represented in Figure 5.16. The mathematical foundation of reinforcement learning is represented by a Markov Decision Process (MDP) (Kamrani et al., 2020). The components are policy (N), Value (V), Actions (A), States (S), and Reward (R).

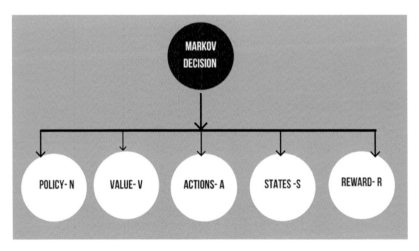

FIGURE 5.16 Markov decision process.

Q-Learning

It is a value-based method of providing information to an agent on what should it take (Clifton and Laber, 2020). Consider the following scenario, depicted in Figure 5.17.

- Consider rooms in a building connected via doors.
- For every room, it is marked with number 0 to 4.
- For the entire building, there is a main door (5)
- All the rooms are connected in such a way that it can lead to the main door 5

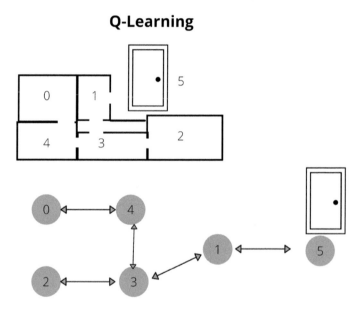

FIGURE 5.17   Q-learning.

Next, start rewarding by the following steps:

- Award a reward of 100 for the doors that lead directly to the goal
- Give reward of 0, when the door does not lead to the target goal
- For two-way doors, double-headed arrows are assigned
- Every arrow in Figure 5.17 holds an instant reward value

Applications of Reinforcement Learning

The applications of reinforcement learning is wider that includes robot manufacturing, business automation, service automation, and so on. It helps in creating training systems that provide custom instructions and materials based on student requirements.

In addition to that, it helps in finding which situation needs an action and also discovers which action will give the highest reward over a period of time. The learning also gives a reward function to the agent and train the agent to get maximum rewards.

Likewise, reinforcement learning cannot be applied in all situations. When there is enough data to solve a problem using supervised learning, reinforcement learning can be omitted. It should also be considered that this learning takes requires more computation and is time-consuming especially when the action space is large.

Reinforcement Learning—Challenges

The major challenges when performing reinforcement learning are as follows:

- It requires great involvement in designing the feature or reward.
- The learning rate is related to the parameters. Fine tuning of the parameters is necessary to speed up the learning process.
- Observability may become partial when it comes to realistic environments and such environments may be nonstationary.
- Results may be diminished when too much of reinforcement is applied leading to overloading of states.

5.3 INCREMENTAL LEARNING FOR BIG DATA STREAMS

Incremental learning is an ML mechanism for streaming data where the learning is a kind of repeated process when new example(s) arise and adjusts the previous learning with the added example(s) (Renuka Devi et al., 2019). In streaming analysis, the assumption of entire availability of data is uncertain before we start the analysis, but the instances may appear over time which is the most significant difference of incremental learning from conventional learning.

Here, input data is constantly used to broaden the existing model's knowledge. It is a dynamic technique of supervised and unsupervised learning that can be applied when the data is available in an incremental manner or its size has out bounded the system memory limits. Such algorithms that facilitate learning incrementally are known as incremental ML learning algorithms.

Generation of streaming data is continuous coming in varied sizes from a wide range of sources, which may include log files generated by buyers using mobile or web applications, e-commerce shopping sites, and social media sources. Such data are unorganized unlike structured data which has a predefined data model. The continuously arriving streaming data requires incremental processing over time and they are rich in volume, variety, veracity, and velocity.

The data is of evolving nature, with large volume and varying time. It is crucial to process such data as they are used in applications such as monitoring and control applications. The data being continuous and to take decisions based on time makes it impossible for processing them after storage. Therefore, the data processing must happen on the go or online. To do so, incremental learning or online learning algorithms will be appropriate for stream data processing. DL will also be beneficial for streaming data as conventional methods are not scalable.

Incremental learning is a family of scalable algorithms that can learn to sequentially renew models from continuous data streams. It does not have all the data available for creating the model unlike the conventional methods. Instead, a living model is built with the data points that come one at a time and the model learns and adapts as the new data comes. The following are the characteristics of an incremental model.

- Prediction can happen at any time.
- It can adapt to concept drift which denotes the changes in the data distribution. For example, to predict and build a model on how much money a bank should loan, it must consider a financial crisis that might change the amount or other factors that. In this case, the model needs to re-learn with a lot of information.
- With finite resources (time and memory), infinite data stream can be processed. It cannot store all the training data unlike conventional ML techniques.

5.4 ENSEMBLE ALGORITHMS

Ensemble techniques are employed where more than one ML model is applied to enhance the results when the problem is complex. The weak models are integrated so that we can accomplish a better model with enhanced accuracy. The ensemble method integrates decision from different ML models and the conclusion is derived.

The weak learner model is also called as "base model." The better choice of model is significant whether it is a classification or a regression. The accurate model is chosen based on the various parameters such as data quantity, features, and dimensionality. The best model is selected with less variance and bias. The model should have bias variance trade-off. The model should have enough flexibility to make the model robust with less variance and bias.

The base models form the foundation of the ensemble; thus combining numerous models together. Most of the base model lacks the good performance because of high bias or variance. The ensemble method is applied not only for better performance also to minimize the variance. The very first step in the ensemble method is to combine all the weak learners.

To integrate the models, we need to identify the base model first. Once a base model is identified then we integrate the rest of the models. Bagging and boosting algorithms are used in general. The base algorithm can be homogenous or heterogeneous. If a single learning algorithm is used for different models, then it is called "homogenous ensemble model," otherwise it is called "heterogeneous ensemble model."

When we aggregate different models together there are two possibilities:

- Model 1: Low bias with high variance
- Model 2: Low variance with high bias

In both aforesaid models, either we aggregate model with reduced variance or aggregate the model with reduced bias. There are different aspects of combining models that are discussed below:

- **Bagging**
  This method is applied for the same type of weak learners and the learning from each of them happen in parallel and the results of the learning are combined by average method.

- **Boosting**
  The learning process is done sequentially, that is, the results of previous decisions are followed up in the next model.

- **Stacking**
  A heterogenous strategy is followed and the learning mechanism is carried out in a parallel mode and the results are combined for final decisions.

The bagging method primarily focuses on creating the ensemble with a reduced amount of variance. The other methods (boosting, stacking) focus on reducing the bias factor. Learning mechanisms can happen in parallel or sequential mode. In parallel mode, learners are executed at the same time and results are aggregated, thus producing the robust model.

The simple ensemble methods are discussed below:

Voting

Voting is the simple method, where each model predicts and votes for a particular output class. This method is generally applied for the classification problem. Each model votes for a particular output class based on the prediction. Based on the votes from each model, majority of the votes are taken for final decision. Consider a classification problem, where the ensemble methods have to classify whether the output class is cancerous or not. The predictions from each ensemble are taken and maximum votes are considered as the final outcome (Figure 5.18).

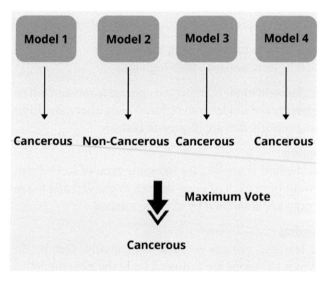

FIGURE 5.18   Voting method.

Average

The average method (Figure 5.19) is very much similar to the voting method but the average of the individual predictions is taken for the final decision. This method is very well suited for regression problems. Consider the below example, where the average of prediction from each ensemble is calculated.

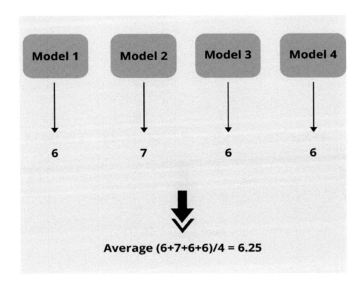

FIGURE 5.19   Average method.

Weighted Average

The average method is extended further into weighted average method. The weights are applied on the individual predictions, then average is calculated. This is shown in Figure 5.20. The weights (0.25) are applied to the individual predictions and average is calculated accordingly.

The advanced methods are bagging, boosting, and stacking. Let us discuss all these methods in detail.

Stacking

The stacking method combines the prediction from different models. Inferences are taken from each prediction to create a new model. Stacking can be applied for generalization, classification, or regression problems. The stacking is done by building more than one base models (base model and meta model). The base model is fit and trained using the training dataset. The meta model is built on the top of the base model predictions. The input to the meta models is obtained from the base model training.

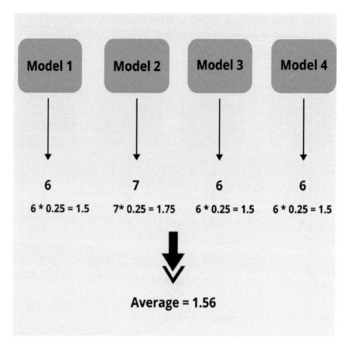

FIGURE 5.20   Weighted average method.

For example, if we consider our base model is KNN, then the training datasets are trained using k-NN and then the predictions are made with the test data. The very same procedure is done for other base models (e.g., decision tree, random forest), and the predictions are made. The final model harnesses the best predictions. The stacking method is depicted in Figure 5.21.

Bagging

Bootstrapping is one of the sampling techniques used to generate subsets from the original datasets. The total size of dataset (N) and the subsets range from s1, s2…,s_n. The total size (N) is equal to the size of the subsets. Bagging method divides the datasets into the number of subsets (Figure 5.22a), to which each model is applied for making predictions. The predictions from each model are collated and combined decision is taken (Figure 5.22b).

The divided subsets are run in parallel to make the predictions. The base model is created for each subset. The individual prediction from each model is collated later.

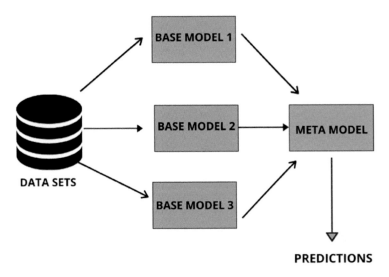

FIGURE 5.21 Stacking method.

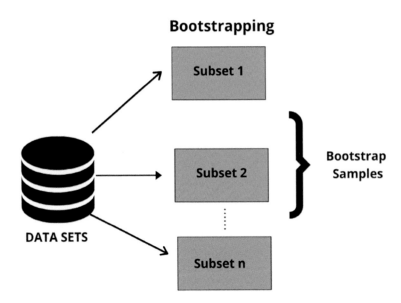

FIGURE 5.22a Bootstrapping method.

We summarize the concept of bagging as:

- Creating subsets from the original dataset

- Fitting models independently and run in parallel (distributed mode)
- Calculating the average of predictions with low variance
- Fitting models iteratively, in sequential mode, based on the training given in the previous step

Bootstrapping

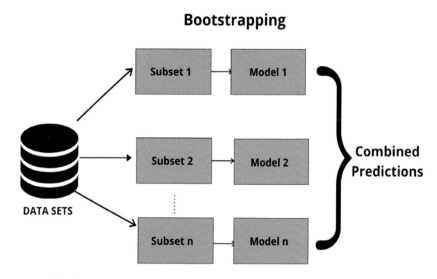

FIGURE 5.22b   Bootstrapping method.

Boosting

Boosting methods are iterative in nature. Several base models are generated iteratively to form a better model with less variance. When the models are generated sequentially, the weak learners follow the adaptive approach (feedback from the previous step and reduces the error). At every iteration, the observations that are missed out in the previous step are given more importance in the next step. Sequentially this process is continued in such a way that all the samples in the datasets are given equal consideration. At the end of iterations, a new model with low variance is built.

The boosting algorithms are further categorized into Adaboost and gradient-boosting algorithms. The algorithms differ in the way the weak learners/models are generated and combined. Each sample in the datasets is assigned with some weights. Adaptive strategy updates the assigned weights in every iteration but the gradient method updates the value of the samples. Both approaches are aimed at creating an optimized model.

The generalized boosting algorithm is given in Figure 5.23.

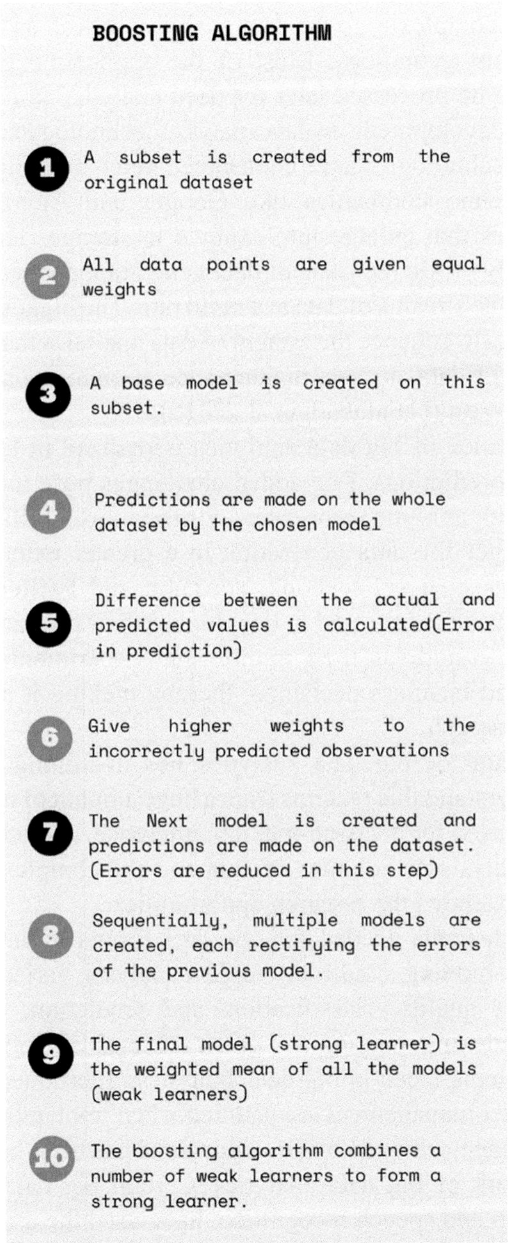

BOOSTING ALGORITHM

1 A subset is created from the original dataset

2 All data points are given equal weights

3 A base model is created on this subset.

4 Predictions are made on the whole dataset by the chosen model

5 Difference between the actual and predicted values is calculated(Error in prediction)

6 Give higher weights to the incorrectly predicted observations

7 The Next model is created and predictions are made on the dataset. (Errors are reduced in this step)

8 Sequentially, multiple models are created, each rectifying the errors of the previous model.

9 The final model (strong learner) is the weighted mean of all the models (weak learners)

10 The boosting algorithm combines a number of weak learners to form a strong learner.

FIGURE 5.23 Boosting algorithm.

5.5 DEEP LEARNING ALGORITHMS

Big data poses complexity to the data analytics as well as the applications that harness this technology. Most of the application gathers massive data that has to be processed later for deep analytics to infer meaningful decisions. The developments in data analytics technology and the increased cloud storage technologies have contributed much in to big data science.

Leading techno companies like Google and Amazon have large data repositories that enlarge into exabyte in storage. The social media organizes hardly tackle the issue of data generation per second, where the users generate the streams of data at a rapid rate. The biggest challenges for these companies to enhance the system of data analytics that accommodate the challenges of data storage, maintenance, monitor, analyze, and other activities of interest (Najafabadi et al., 2015).

The importance of big data analytics is realized in term of inferred decisions and predictions. The added challenges pose more complexity to the ML and analytic processes. Furthermore, billions of people constantly trigger the data generation to a greater extent in the social media platforms such as Facebook, YouTube, and Twitter. A wide range of organizations have invested in developing their products utilizing big data analytics to address their monitoring, experimentation, analysis, simulations, and business decisions, thereby making it a focal point in data science research.

The main aim of big data analytics lies in mining and extracting significant understandable patterns from a huge amount of input data which will be further used for decision-making, inference, and other predictions. Along with analysis of volumes of data and the complexity of big data streams have extended the research opportunities.

The research focus needed for raw data format variation, streaming data that is fast-moving, credibility of data analysis, varied input sources, noisy and data quality, classification, and prediction. Sufficient data storage, data indexing, tagging and quick retrieval of the information are the major problems faced in big data analytics. Subsequently, novel data analysis and data management are justified when implementing big data.

The knowledge achieved by DL algorithms has been basically unused in the framework of big data analytics. Certain big data domains, like computer vision and speech recognition, have used the application of DL majorly to develop the classification modeling results.

The capability of DL algorithms to extract advanced, complex abstractions and representations of huge volumes of data particularly unsupervised makes it a smart and important tool for big data analytics. Particularly, big data problems like data tagging, semantic indexing, quick retrieval of information and data modeling can be clearly approached and solved with help of DL.

Deep learning includes:

1. fairly simple linear models that can work efficiently with the knowledge received from the more complex and more conceptual data representations,
2. high increase in automation of extraction of data representation from unsupervised data makes its broad application to various data types, such as text, image, audio, etc., and
3. the semantic and relational knowledge can be achieved at the advanced levels of abstraction and representation of the raw data.

DL architectures are more suitable to address the issues pertaining to the characteristics of big data analytics. DL fundamentally exploits the accessibility of huge quantities of data, that is, Volume, where the algorithms with narrow learning hierarchies flop to discover and recognize the advanced complexities of data patterns.

Furthermore, since DL compacts with data abstraction and representations, it is quite expected to be suitable for analyzing raw data that is presented in various formats and/or from various sources, that is, Variety, and this may lessen the need for human experts giving input to extract features from the existing data.

While offering various challenges for further conventional data analysis methods, big data analytics grants significant opportunity for evolving novel procedures and models to address issues related to big data in particular. One such solution for data analytics experts and practitioners is provided by DL.

A prime task involved with big data analytics is information retrieval. Well-organized storage and data retrieval is a rising problem in big data, particularly since massive quantities of data such as text, audio, video, and image are being collected and made readily available across various domains, like social networks, marketing systems, shopping and security systems, defence systems, fraud detection, and cyber traffic monitoring. Earlier strategies and solutions used for data storage and retrieval are challenged by the large volumes of data and various data formats, linked to big data.

Handling with streaming and fast-moving input data is one of the major challenges encountered in big data analytics. It is practically useful in supervising tasks such as fraud detection. It is of prime importance to adapt DL to have a grip on streaming data as it need to have algorithms that can handle with massive amounts of input data which are continuous. This section discusses some of the mechanisms associated with DL and data streaming inclusive of incremental learning and classification techniques.

The main advantage of DL is it provides potential solution to address the problem of data analysis and learning with large amount of input data. More particularly, it helps to extract complex data representation from huge amount of unsupervised data. Thus, DL pertains to big data analysis from massive assortments of data that is usually unsupervised and un-categorized.

The advantage of DL is the ability to explore the complexity levels in a hierarchical way thus making the analytics task as a simpler procedure. This will be beneficial when analyzing large data, to index data based on semantics, tagging, and information retrieval.

The hierarchical learning and extraction of different levels of complex, data abstractions in DL paves a way to simplification for big data analytics tasks, especially for analyzing huge amount of data, semantic indexing, data tagging, information retrieval, and classification. Figure 5.24 gives a clear understanding of the relationship between AI, ML, and DL.

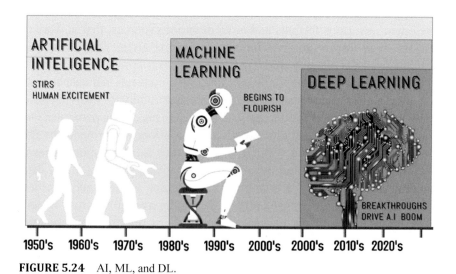

FIGURE 5.24   AI, ML, and DL.

DL has been used productively in a lot of applications, and is measured to be one of the most progressive ML and AI techniques currently. The algorithms associated with it are often used for unsupervised, semisupervised, and supervised learning problems. The number of layers is greater in case of the DL models based on neural network when compared with the shallow algorithms.

Shallow algorithms likely to be less complex and necessitates more knowledge of most favorable features to use, which in general applied for feature selection and engineering. On contrary, DL algorithms depend on optimal model in the course of model tuning. These algorithms are apt in solving a problem that lacks the knowledge of features when it is unlabeled or not essential for the primary use case.

The different DL model architectures and learning algorithms include:

- Feedforward neural networks
- Recurrent neural network
- Multilayer perceptrons (MLP)
- Convolutional neural networks
- Recursive neural networks
- Deep belief networks
- Convolutional deep belief networks
- Self-organizing maps
- Deep Boltzmann machines
- Stacked denoising autoencoders

5.6 DEEP NEURAL NETWORKS

The interior working nature of a human brain is often modeled as a concept of neurons and the connected neurons are called as biological neural networks. As per Wikipedia, it is roughly calculated that human brain contains 100 billion neurons which are connected through these networks.

Artificial neural networks (ANNs) are directly inspired by and partly modeled based on biological neural networks. They are competent of modeling and processing the nonlinear relationships between parallel inputs and outputs. The algorithms related to it are part of the wide range of ML and can span many applications.

ANNs comprise of adaptive weights that are assigned between neurons. The learning algorithm can alter the weights from the experience in order to progress the model. The cost function selection is vital in adjusting the weights along with the learning algorithm and providing the optimal solution to the problem.

This includes defining the best values for all of the model parameters that are tuneable, with the primary target being neuron path adaptive weights, along with tuning parameters algorithm such as the learning rate. The parameter tuning is achieved by gradient descent or stochastic gradient descent techniques. These optimization methods basically attempt to bring the optimal solution, once it is reached that optimal level the model is ready to solve a given problem.

ANN is architecturally modeled using layers of artificial neurons that receive a input and pass the messages based on the threshold value (activation function). A simpler model comprises of input, one or more hidden, and an output layer (Figure 5.25).

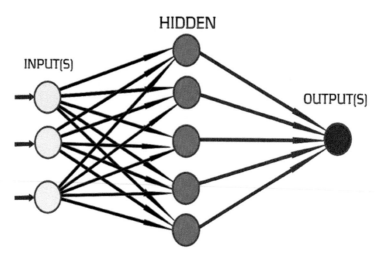

FIGURE 5.25 Artificial neural networks.

Every layer has one or more neurons. Models can become highly complex, and with amplified abstraction and problem-solving capabilities with added layers, and connectivity between the neurons.

Let us look at what a perceptron means. The dendrites are extensions of the nerve cell as given in the neuron cell in Figure 5.26. They are in charge of receiving the signals and then transmitting the signals to the body cells, which will process the stimulus and takes a decision to transmit the signals to other neuron cells. If it decides to trigger signals, the axon will trigger the chemical transmission to other cells.

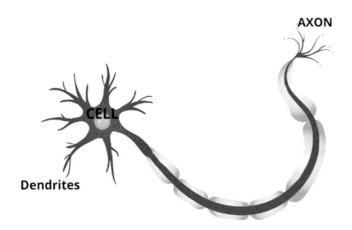

FIGURE 5.26 Neuron.

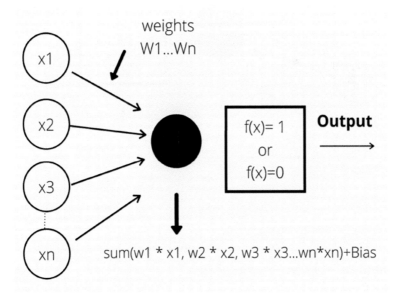

FIGURE 5.27 Perceptron process.

The procedure of processing of data by perceptron is as follows:

1. $x_1, x_2, x_3, \ldots, x_n$ are input neurons.
2. $w_1, w_2, \ldots, w_m$ are weights, applied to the neurons $w1 * x_1$, $w_2 * x_2$, $w_3 * x_3$ and so on.
3. The weights are multiplied with the input and bias is added further.
4. The function $f(x)$ yields either 0 or 1, based on the step 3 calculation. When the calculated sum is larger than 0, then the output is 1 otherwise it is 0.
5. The outcome from the network $f(x)= 1$ or 0.

The process is represented as follows with an enhanced diagram in Figure 5.28.

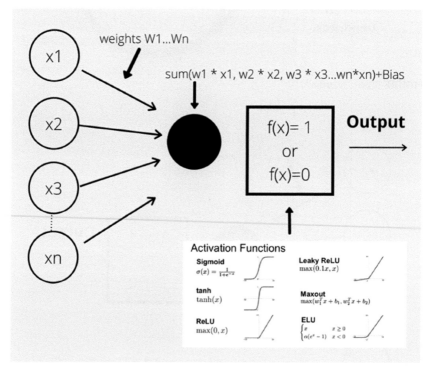

FIGURE 5.28 Enhanced perceptron process.

1. Perceptron gets the input
2. Inputs are multiplied with weights

3. Bias is added with a multiplied sum.
4. Then, activation function (sigmoid, tanh, relu) is applied.
5. The result is either 1 or 0.

Training Neural Networks

The process of training the neural networks engages utilizing an optimization algorithm to find a set of weighs to map the inputs to outputs. A bigger weight indicates a tight correlation between a signal and the network's outcome.

The network interpretation of the data will be affected more when the inputs are paired with large weights rather than having the inputs paired with smaller weights. This procedure of learning using weights for any learning algorithm is the process of readjusting the weights and biases which allocates importance to certain bits of information and minimizes other bits.

This model helps us to understand which predictors are connected to which outcomes and make the weights and biases accordingly. The activation function with and without bias is given in Figure 5.29. Bias permits you to change the activation function by a constant value and it merely acts like a linear function, where the line is efficiently transposed by the constant value.

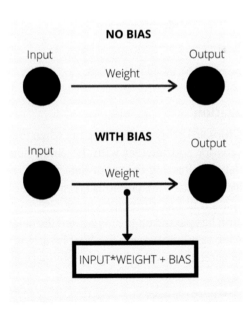

FIGURE 5.29 NN with bias.

The network is trained with weights and bias parameter. The iterative learning is processed by the layers for better learning. In forward propagation, learning proceeds in a forward direction. The backpropagation happens in a reverse direction to reduce the error rate by adjusting weights and activation function. This is shown in Figure 5.30.

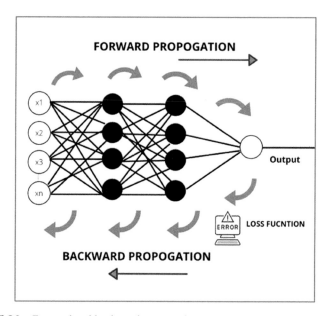

FIGURE 5.30   Forward and backward propagation.

Forward Propagation

In the first phase, forward propagation happens when the training data gets exposed to the network and interact with the entire network to calculate the predictions (labels). Passing the input data via the network in such a manner that all the neurons of the previous layer transform the information they receive and pass it to the subsequent layer.

When the neuron has passed all the layers and the calculations are made by all its neurons the result will reach the final layer with label prediction. Loss function measures the level of error by comparing the actual with a prediction.

Preferably, we require reduced cost function value to maintain the deviation among estimated and expected value. Consequently, as the model is being trained, the parameters are tuned until good predictions are obtained.

Backpropagation

The information is propagated backward once the loss has been estimated. Therefore, its name: backpropagation. Beginning with the output layer, that loss information broadcasts to all the neurons in the hidden layer that contributes directly to the output.

Still, the neurons present in the hidden layer receives only a part of the signal loss, based on the contribution relative to that of the original output contributed by each neuron. This process is repetitive, at each layer until all the neurons present in the network have suffered a loss signal that contributes to the total loss.

The aim is at making the loss down to zero, when we train the network for a prediction.

To reduce and adjust the loss, we use gradient descent. This method tries to adjust the weight by changing into small variations to reduce the loss and project toward the global minimum. The mechanism is repeated for a number of epochs to cover training for entire dataset.

The learning algorithm works as follows (Figure 5.31):

- It starts with the weights (w_{ij}) and biases (b_j).
- Fed the input into the network to get the prediction.
- Loss is found by comparing the predicted with the actual values.
- To overcome the loss, backpropagation is performed to fine tune the parameters.
- When we adjust the parameter that paves a way for an enhanced model with less error.
- The previous steps are repeated until better accuracy of the model is achieved.

Activation Functions

Linear

One of the activation functions is Linear which is fundamentally an identity function, that is, the signal does not change (Figure 5.32).

Sigmoid

The sigmoid activation function is used to convert independent variables which are almost infinite in range into simple probabilities of values between 0 and 1 (Figure 5.33). Mostly the output is close to the extremes of values 0 or 1.

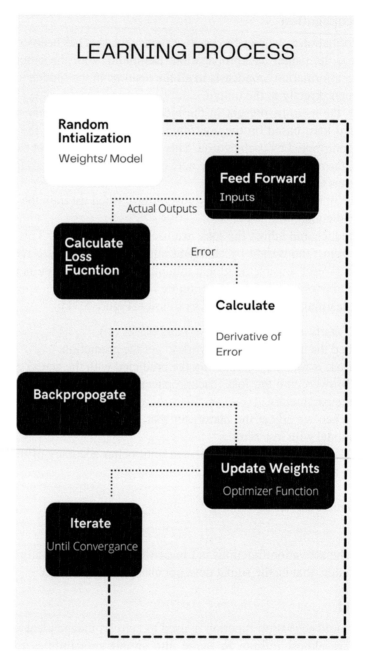

FIGURE 5.31   Learning process.

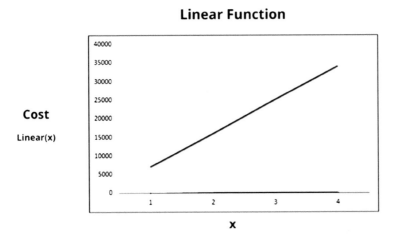

FIGURE 5.32   Linear function.

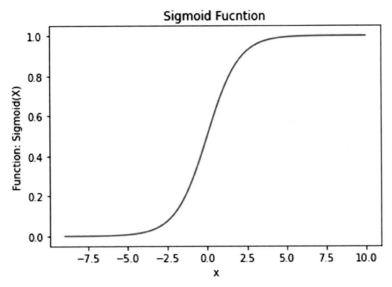

FIGURE 5.33   Sigmoid function.

Tanh

The "tanh" activation function is used to represent the hyperbolic sine and hyperbolic cosine relationship

$$tanh(x)=sinh(x)/cosh(x).$$

The range of "tanh" ranges between −1 and 1, this is the input that adheres with some neural networks. The major advantage of "tanh" is that it can easily deal with negative numbers. It is represented in Figure 5.34.

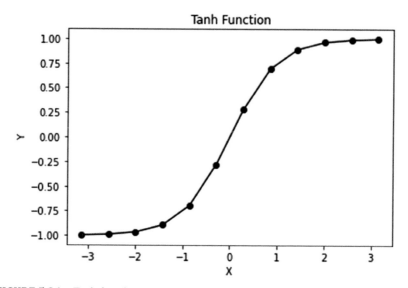

FIGURE 5.34   Tanh function.

Softmax

The softmax activation function can be useful for more than one output values (for each label). This function is mostly found on the neural network's outer layer and the values it produces are the probability distribution over multiple vectors (Figure 5.35).

ReLU

The rectified linear unit (ReLU) activation function commonly referred to as ReLU is an interesting approach in which if an input is greater than certain threshold value a single node is activated. The most common behavior of this activation function is the output remains as zero as long as the input is zero.

When the input raises, the output is in linear relationship with the input value, which is of the form $f(x) = x$. This activation function has been verified to work in many diverse situations and is currently extensively used. It is represented in Figure 5.36.

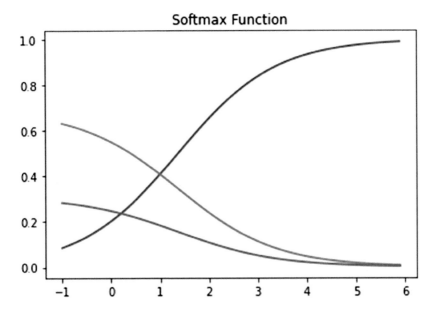

FIGURE 5.35   Softmax function.

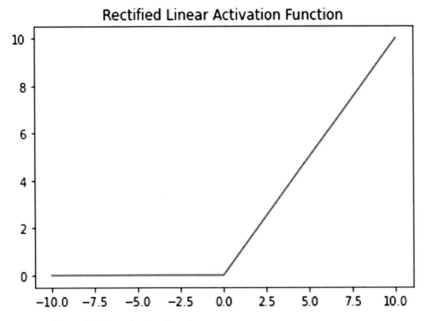

FIGURE 5.36   ReLU activation function.

The categories of DL algorithms are discussed below,

5.7 CATEGORIES OF DEEP LEARNING ALGORITHMS

1. Convolutional Neural Networks (CNNs)
2. Long Short-Term Memory Networks (LSTMs)
3. Recurrent Neural Networks (RNNs)
4. Generative Adversarial Networks (GANs)
5. Radial Basis Function Networks (RBFNs)
6. Multilayer Perceptrons (MLPs)
7. Self-Organizing Maps (SOMs)
8. Deep Belief Networks (DBNs)
9. Restricted Boltzmann Machines (RBMs)
10. Autoencoders

The DL algorithms mentioned above are used to solve complex issues with data of any kind. These algorithms require huge amount of computing power along with information. Now let us take a closer look on these algorithms.

5.7.1 CONVOLUTIONAL NEURAL NETWORKS

Convolutional neural networks (CNNs), also well-known as ConvNets, consist of manifold layers and are primarily implemented for object detection and image processing. It was first developed in the year 1988 by Yann LeCun. It was called LeNet and was used for recognizing ZIP codes and digits. CNNs are extensively used to spot satellite images, forecast time series, process medical images, and sense anomalies.

Procedure

CNNs have numerous layers that process and mine features from data:

Convolution Layer

- This layer has a number of filters to carry out the convolution operation, which excerpts the features from the given image. It takes the image matrix as the input along with a filter. The results are output vector after extracting the features.

Rectified Linear Unit

*   CNNs contain a ReLU layer to execute operations on elements. If the signal is negative then its output is zero, for positive value it returns the same. The feature map is produced.

Pooling Layer

*   The obtained rectified feature map is next fed into a pooling layer. Pooling is a process of downsampling that minimizes the feature map size.
*   The formed 2D feature map is converted by the pooling layer into a single, stretched, continuous, and linear vector.

Fully Connected Layer

*   The flattened matrix obtained from the previous layer when fed as input forms into a fully connected layer which categorizes and identifies the images.

Example of an image processed through CNN is presented in Figure 5.37.

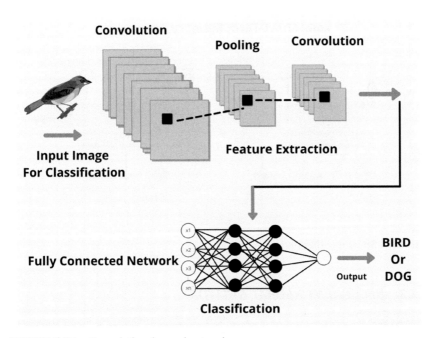

FIGURE 5.37 Convolutional neural networks.

5.7.2 LONG SHORT-TERM MEMORY NETWORKS

Long short-term memory networks (LSTMs) are a category of RNN that can be trained and remember long-term dependencies. Recalling the past information for elongated durations is the default behavior. LSTMs hold information over long period of time. They come in handy for time-series prediction since they recollect earlier inputs. LSTMs is a connected structure.

The structure contains four interacting layers which communicate in an exclusive way. In addition to time-series predictions, LSTMs are usually implemented for speech recognition, pharmaceutical development, and music composition.

Procedure

- First, they ignore and do not remember the unrelated states.
- Then, particular cell-state values are updated.
- At the end, output the cell state.

Working of LSTMs is presented in Figure 5.38.

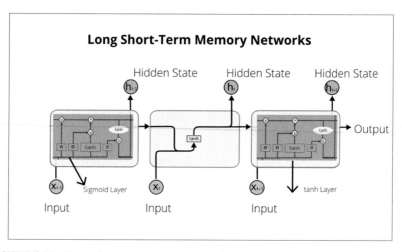

FIGURE 5.38   Long short-term memory networks.

5.7.3 RECURRENT NEURAL NETWORKS

The RNNs have the connections forming direct cycles, thereby allowing the outputs of LSTM to be passed as inputs to the existing phase. The output from the LSTM is passed as an input to the existing phase and can keep track of its previous inputs due to its internal memory.

RNNs are commonly applied for, time-series analysis, image captioning, natural-language processing, handwriting recognition, and machine translation. An unfolded RNN looks like this. An unfolded RNN is presented in Figure 5.39.

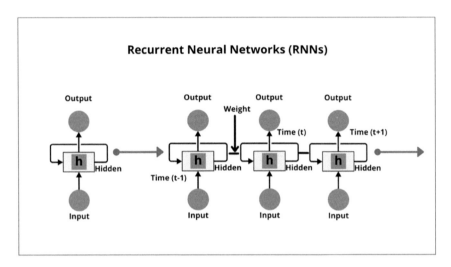

FIGURE 5.39 RNNs.

Procedure

- The input arrives at a time t, with the output rate $t-1$.
- Output time (t) is again fed into the input time (t).
- It can process varied lengths of input.
- Size of the model does not depend on the input size.

Figure 5.40 shows a sample of how Google's autocompleting feature works.

5.7.4 GENERATIVE ADVERSARIAL NETWORKS

GANs are generative DL technique that creates new data instances from the given samples. The main components of this architecture are generator and discriminator. The functionality of the generator is to duplicate the given samples, wherein the discriminator learns and identifies the fake data.

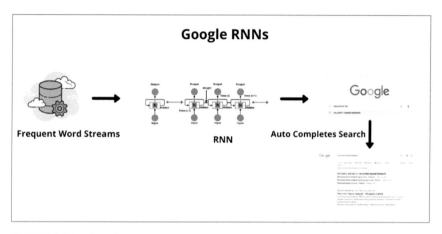

FIGURE 5.40   Google RNNs.

The usage of GANs has amplified over a period of time. The applications include photo realistic images, image editing, 3D object generation, and attention specific predictions in image, cartoon characters, and so on. Specifically, in game development to convert the low dimensional images into higher resolutions.

Procedure

- The discriminator identifies the fake data from the real data.
- During the learning process (training) the generator introduces the fake data. The discriminator trains to locate the fake from real.
- The results of the training were updated with the model. Thus, generator results are synced with the discriminator.

Figure 5.41 shows the GANs procedure.

5.7.5   RADIAL BASIS FUNCTION NETWORKS

RBFNs apply the radial basis function. The architecture has input, hidden, and output layer. RBFNs are mostly applied in areas for classification, regression, and time-series prediction. Figure 5.42 represents RBFN.

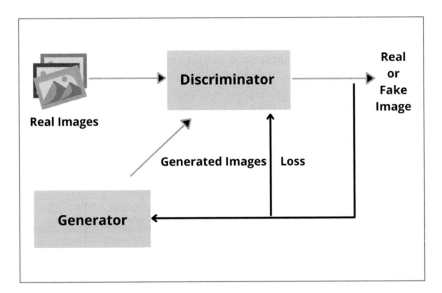

FIGURE 5.41 Generative adversarial netwroks.

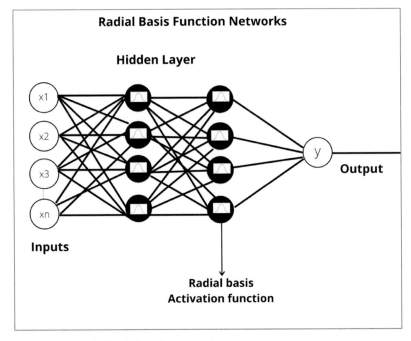

FIGURE 5.42 Radial basis function networks.

Procedure

- This network classifies the given input based on the similarity measures learned during the training process.
- Every layer has input and output vector, with the neurons.
- The activation function is applied and the weighted sum of the input is calculated for each category of data.
- The hidden layer neurons are associated with Gaussian function
- The network produces the output which is the linear combination of given inputs

5.7.6 MULTILAYER PERCEPTRONS

MLPs are the best to clearly learn about DL technology. MLPs are constructed as a feedforward network of connected layers of perceptrons. MLPs have an input layer and an output layer that are fully connected. They have the equal number of input and output layers but it can extend with many hidden layers that can be used to build software for image, speech, and machine translation.

Procedure

- The data is fed into the input layer. The layers enable the data to move forward into the next level of layers.
- The weights are calculated for each input.
- Activation functions are applied on nodes.
- Train the model—The learning mechanism enables the network to know the similarities and correlations between the dependent and independent variables.

Figure 5.43 is an example of an MLP. The diagram shows classification of cat and dog images.

5.7.7 SELF-ORGANIZING MAPS

Developed by Professor Teuvo Kohonen, SOMs enable data visualization, which is used to reduce the dimensions of data by using self-organizing artificial neural networks. The data visualization tries to solve the high-dimensional data problem that a human being cannot easily visualize.

SOMs are developed to help users understand this high-dimensional information.

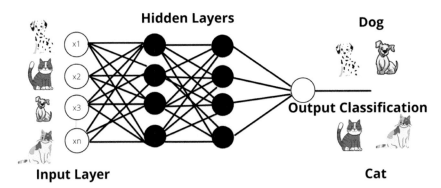

FIGURE 5.43 Multilayer perceptrons.

Procedure

- Weights are initialized for each input vector, and the vector is chosen at random during training process.
- Examines each node to find the suitable vector weights based on the input vector. The selected node is Best Matching Unit (BMU).
- The neighborhood of BMUs is explored.
- Awards a winning weight. When a node is closer to BMU, the weight changes accordingly.
- The steps are repeated for number of times.

Figure 5.44 shows an input vector of different colors. This data is feed to an SOM that converts into 2D RGB and categorizes in different colors.

5.7.8 DEEP BELIEF NETWORKS

DBNs are generative models and it is a stack of restricted Boltzmann machines. The RBM establishes the connectivity between the top and bottom layers. The end of the stack is softmax layer that creates a classifier. Every layer has the dual role as the input as well as the hidden layer. DBNs are applied in areas for image-recognition, video-recognition, and motion-capturing data. DBN architecture is given in Figure 5.45.

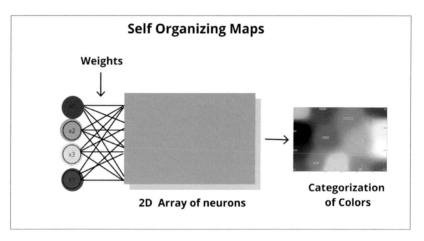

FIGURE 5.44   Self-organizing maps.

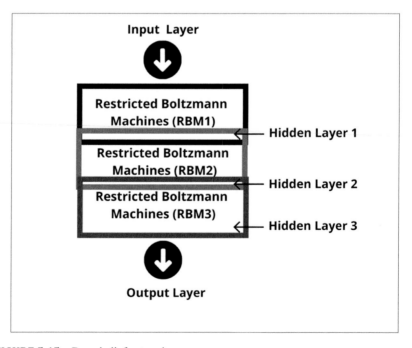

FIGURE 5.45   Deep belief networks.

Procedure
- Greedy learning algorithms are used.
- The layer-by -layer method is employed to produce the weights by top-down approach.
- Gibbs sampling is the statistical method based on the Bayesian theory. The sampling method is applied on the top hidden layers.
- Samples are drawn in a single pass of sampling.
- Until the convergence, the training is done for a number of iterations.

5.7.9 RESTRICTED BOLTZMANN MACHINES

RBMs are a stochastic neural network that learns from a probability distribution over a set of inputs. It was developed by Geoffrey Hinton. This DL algorithm is applied in areas for dimensionality reduction, classification, regression, and feature extraction. RBMs consist of two layers namely visible and hidden units, where both are connected to each other.

Procedure
- Has forward and backward pass mechanism.
- Each input is assigned with weights and activation function, bias is applied when it is moved from the input layer to the hidden layer.
- During backpropagation, the set of numbers are translated into number of inputs.
- Individual weights along with bias are added and the output is passed into a visible layer.
- Finally, the output is compared with the input to have a quality check.

Figure 5.46 shows how RBMs function.

5.7.10 AUTOENCODERS

Autoencoders are a particular kind of feedforward neural network in which the input and output are identical. It was designed by Geoffrey Hinton in the 1980s to solve the unsupervised learning problems. The autoencoders are trained neural networks that imitate the data from the input layer to the output layer. Autoencoders are applied for purposes such as popularity prediction, pharmaceutical discovery, and image processing.

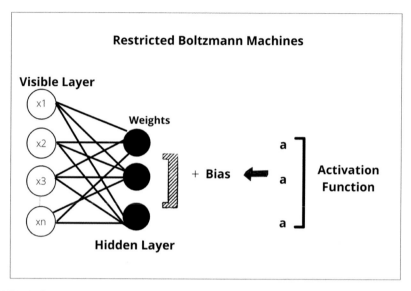

FIGURE 5.46　DBNs.

Procedure

An autoencoder involves main mechanisms such as encoder, code, and decoder.

- Components are the code, encoder, and decoder.
- The functionality of the autoencoders to alter the given input into another form of representation. They are trained to recreate the original form of the input.
- When the image clarity is degraded or not visible, the auto encoder is used to code them.
- The first process of the autoencoder is to encode the given image into a smaller size image.
- Then, the next step is to decode the image into a larger size.

Figure 5.47 demonstrates autoencoders.

5.8　APPLICATION OF DL-BIG DATA RESEARCH

DL algorithms are advantageous in understanding the semantics of the data representations that make easy retrieval of information. Searching and indexing are the core functions of language processing. The raw data are processed by DL algorithms and converted into the corresponding vector representations, thus makes the comparison process easier instead of processing the raw data.

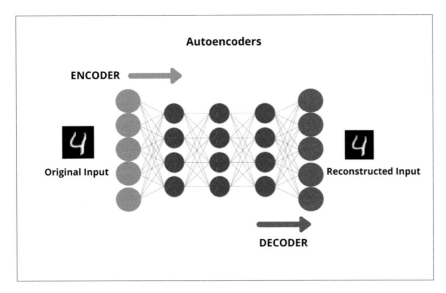

FIGURE 5.47 Autoencoders.

The "Word2Vec" tool extracts the semantics from the big data. The extracted vector representations are used for search and retrieval. The tool converts the large texts (input) into vector representations, which are the basics for NLP and ML applications. The semantic indexing is the fastest technique of information retrieval used for big data analytics text processing. The search engines adapt this mechanism for storage and retrieval.

DL methods are also employed for discriminative analytics, for extracting the nonlinear features from the given data. Extracted features are applied to the linear models for further analytics, thus discriminative analysis is used for image and video analysis and compression. Semantic tagging is a kind of application where DL algorithms are used. Semantic tagging is used to associate a document with the relevant sources. DL can also be applied for video tagging and object recognition.

DL algorithm's feature extraction capability is used in many image segmentation and classification problems. When the data is very complex, feature extraction is a challenging task. The research is further focused on exploring the different data representations and complex issues.

With the development of CNN architecture, large-scale image segmentation and semantic segmentation are made possible with reduced time complexity and scalability. CNN is specifically used for large-scale image classification, where feature extractions are very rapid.

Industrial automation can be improved by leveraging DL-based big data technologies. The centralized production process and supply chain management are the core functions of any industry. The other operations include assembling data from sensors, scrutinizing the time series data, sales prediction, customer sentiment analytics, and product segmentation.

For a successful business customer feedback, usage analytics, their sentiments toward the product are crucial factors. Sentiments of the people is an important factor for business. The positive, negative, or neutral sentiments are vital for company's reputation; so analyzing these sentiments gathered by social media is really important. User sentiments, twitter streams are valuable sources of big data. These kinds of behavioral analytics leverage the business activity to a greater extent.

DL-based systems enhance the business opportunity as well as sales. Social media analysis is a complex task. DL coupled with NLP systems is in use for analyzing the user sentiments represented in the natural language. Many research facets have been proposed in social big data analytics.

In the present scenario, Internet security and cybersecurity are the need of the hour. The intrusion detection system uses DL algorithms to combat all sorts of network attacks and identifies the unknown attacks that are not exposed under the conventional network management systems. The enhanced capability of the system identifies and tackles all the network-based issues in real time thus providing on time real-time solution.

DL-based urban analytics unlock the possibilities of urban development with deep analytics. The fusion of this technology leads to urban computing which learns the features and explores the feature representation for classification problems. The analytics are useful in urban planning, traffic, environment, energy consumption, and public security.

KEYWORDS

- **machine learning**
- **deep learning**
- **CNN**
- **ANN**
- **supervised learning**
- **unsupervised learning**
- **reinforcement learning**

REFERENCES

Aggarwal, C. C. Neural Networks and Deep Learning. *Springer* **2018**, *10*, 978–983.

Boser, B. E.; Guyon, I. M.; Vapnik, V. N. A Training Algorithm for Optimal Margin Classifiers. In *Proceedings of the Fifth Annual Workshop on Computational Learning Theory—COLT '92*, 1992; p 144.

Clifton, J.; Laber, E. Q-learning: Theory and Applications. *Annu. Rev. Statist. Its App.* **2020**, *7*, 279–301.

Cunningham, P.; Delany, S. J. *k-Nearest Neighbour Classifiers*. arXiv preprint arXiv: 2004.04523, 2020.

https://en.wikipedia.org/wiki/Logistic_regression#:~:text=Logistic%20regression%20 is%20a%20statistical,a%20form%20of%20binary%20regression)

https://en.wikipedia.org/wiki/Machine_learning

https://machinelearningmastery.com/linear-regression-for-machine-learning/

https://machinelearningmastery.com/types-of-learning-in-machine-learning/

https://medium.com/analytics-vidhya/machine-learning-development-life-cycle-dfe88c44222e

https://medium.com/datathings/neural-networks-and-backpropagation-explained-in-a-simple-way-f540a3611f5e

https://towardsdatascience.com/ensemble-methods-bagging-boosting-and-stacking-c9214a10a205

https://towardsdatascience.com/how-does-linear-regression-actually-work-3297021970dd

https://towardsdatascience.com/introducing-deep-learning-and-neural-networks-deep-learning-for-rookies-1-bd68f9cf5883

https://towardsdatascience.com/spam-detection-with-logistic-regression-23e3709e522

https://www.analyticsvidhya.com/blog/2018/06/comprehensive-guide-for-ensemble-models/

https://www.guru99.com/data-mining-tutorial.html\

https://www.guru99.com/images/1/082319_0514_Reinforceme4.png

https://www.ibm.com/cloud/learn/supervised-learning

https://www.innoarchitech.com/blog/artificial-intelligence-deep-learning-neural-networks-explained

https://www.kdnuggets.com/2019/10/introduction-artificial-neural-networks.html

https://www.pico.net/kb/the-role-of-bias-in-neural-networks

https://www.simplilearn.com/tutorials/deep-learning-tutorial/deep-learning-algorithm

Jan, B.; Farman, H.; Khan, M.; Imran, M.; Islam, I.U.; Ahmad, A.; Ali, S.; Jeon, G. Deep Learning in Big Data Analytics: A Comparative Study. *Comput. Electr. Eng.* **2019,** *75,* 275–287.

Kamrani, M.; Srinivasan, A. R.; Chakraborty, S.; Khattak, A. J. Applying Markov Decision Process to Understand Driving Decisions Using Basic Safety Messages Data. *Trans. Res. Part C: Emerg. Technol.* **2020,** *115,* 102642.

Kelleher, J. D.; Namee, B. M.; D'arcy, A. *Fundamentals of Machine Learning for Predictive Data Analytics: Algorithms, Worked Examples, and Case Studies*; MIT Press, 2020.

Liu, J.; Li, T.; Xie, P.; Du, S.; Teng, F.; Yang, X. Urban Big Data Fusion Based on Deep Learning: An Overview. *Inf. Fusion* **2020,** *53,* 23–133.

Najafabadi, M. M.; Villanustre, F.; Khoshgoftaar, T. M.; Seliya, N.; Wald, R.; Muharemagic, E. Deep Learning Applications and Challenges in Big Data Analytics. *J. big Data* **2015,** *2* (1), 1–21.

Renuka Devi, D.; Sasikala, S. Online Feature Selection (OFS) with Accelerated Bat Algorithm (ABA) and Ensemble Incremental Deep Multiple Layer Perceptron (EIDMLP) for Big Data Streams. *J. Big Data* **2019,** *6,* 103.

Saxena, A.; Prasad, M.; Gupta, A.; Bharill, N.; Patel, O. P.; Tiwari, A.; Er, M. J.; Ding, W.; Lin, C. T. A Review of Clustering Techniques and Developments. *Neurocomputing* **2017,** *267,* 664–681.

Xu, S. Bayesian Naïve Bayes Classifiers to Text Classification. *J. Inf. Sci.* **2018,** *44* (1), 48–59.

CHAPTER 6

Case Study

This chapter highlights application across verticals with research problems and solutions. Healthcare analytics is characterized as quantitative and qualitative processes used to improve healthcare efficiency by computers, servers, or cloud-based applications that store and categorize data and draw conclusions. Effective healthcare analytics provide organizational intelligence that helps to explore health data. Various big data analytic methods for health care analytics are discussed with case study. The application of big data in social media is called as social media analytics.

Many companies have started using this platform for business in particular social media advertising has been increased. Businesses are adopting the power of data and technology in order to become more impartial and data-driven. Big data analytics is important because the mostly unstructured data, which accounts for nearly 80% of all data, can help companies uncover hidden opportunities, better understand their clients, and optimize business processes.

6.1 INTRODUCTION

Big data is a concept of massive amounts of structured and unstructured data that overwhelm any organization's day-to-day operations. Its advanced analytical techniques are used to interpret, analyze, and extract information from the complex datasets that are stored on the servers. The informatics derived from this technique will be greatly helpful in generating multiple findings for extensive studies. Moreover, big data offers vast opportunities

Research Practitioner's Handbook on Big Data Analytics. S. Sasikala, PhD, D. Renuka Devi, & Raghvendra Kumar, PhD (Editor)
© 2023 Apple Academic Press, Inc. Co-published with CRC Press (Taylor & Francis)

for varied aspects of data research that can fetch desired results and predictions with low latency.

Predictive modeling and analytics play a critical role in improving the overall quality of patient care in healthcare. Pharmaceutical corporations, insurance companies, public health officials, physicians/surgeons, hospitals, and patients all have a keen interest in analytics in the healthcare industry.

It should be obvious that the quantity of data in a company does not matter; what matters is what the company does with the data. Big data can be interpreted for insight, which contributes to improved decision-making, better business practices, and, most importantly, the preparation of strategic business steps that support both companies and customers.

The data that was already on the servers was data that had been sorted and filed up until yesterday. Big data has recently become standardized and universally recognized as a slang term. The term refers to any amount of data that a company owns that is still current. It includes data from clouds, data centers, and even personal information about employees.

Big social data is highly unstructured, voluminous, and semantic in nature. The sources of data are gathered from the logs of Twitter, Facebook, and other social media interactions. By analyzing and establishing patterns from these kinds of interactions will lead to analyze their behaviors and social interests.

Advanced big data analytics can assist higher-education executives to engage existing and future students, raise student enrolment, improve student retention, and completion rates, and even improve teacher productivity and study. There are many different types of data analytics that can be used to create reports and dashboards, ranging from simple to complex:

- *Exploratory data analysis*—This approach is used to find data patterns and connections.
- *Confirmatory data analysis*—This approach employs mathematical methods to determine whether data hypotheses are accurate.
- *Data mining*—In this approach, to detect trends or patterns, it is necessary to sift through vast amounts of data.
- *Predictive analytics*—This approach makes use of data to attempt to forecast consumer, facilities, or other incident actions.
- *Machine learning*—This employs AI (Artificial Intelligence) and pre-programmed algorithms to look through data in the place of a person.

- *Business Intelligence (BI)*—It is a collection of processes and technology that transform raw data into usable knowledge in order to improve decision-making.

In this chapter, we look at a few big data analytics case studies w.r.t. healthcare, business, social, and educational domain.

6.2 HEALTHCARE ANALYTICS—OVERVIEW

Healthcare is an excellent illustration of how big data analysis tools are being used to their full potential. Data about healthcare come from systems, medical records, clinicians, and hospitals. The generated data are primarily used by the government, researchers, automated systems, and so on. This data accumulation acts as a repository across the globe for data, thus providing data transparency.

Even though the complexity of healthcare data is additional, the advantage in handling and maintaining the voluminous data by implementing big data techniques is the need of the hour. According to McKinsey Global Institute, US healthcare might benefit from big data approaches, which may generate $300 billion in income each year.

The conventional approach to the healthcare sector generally focuses upon the investigation of states of diseases and observing the changes that happen in physiology. The conventional approach lacks the better understanding of diagnosing the diseases and arriving at the conclusions. The huge volume of data was gathered but none of the methods were available to utilize these data for the betterment of treatment.

After the rapid technological boom, greater transformation in the dimension of approaching and handling patients' healthcare has surfaced across the globe. With the emerging technology, every aspect in the healthcare sector is now transposed to better inferences yielding reduced time to treat patients. In recent times, big data has facilitated clinical assessments that are made by analyzing different measures like heart rate, blood pressure, respiration count, oxygen level, and so on.

When we employ big data technology to assimilate all these factors together, we can create the possibility of clinical assessment in a more convenient and competent way. Both structured and unstructured data are obtained and used for complete disease analytics in this integrated method. Healthcare big analytics research has gained much interest amongst researchers. Most of the works are related to handling the underlying streaming characteristics of data and establishing the taxonomy of analytics.

6.2.1   HEALTHCARE ANALYTICS—BENEFITS

Healthcare analytics is defined as the quantitative and qualitative procedures carried out by computers, servers, or cloud-based apps to enhance healthcare efficiency by storing and categorizing data and drawing conclusions based on emerging trends.

In a company, the historical data, age-old records, and new information from internal or external sources, all these datasets can be simultaneously analyzed and also be subjected into comparative studies. Data analysis in the healthcare industry can help businesses increase revenue, improve efficiency, improve customer service, and devise plans ahead of competitors in the marketplace.

The application of the aforementioned methods in the healthcare industry is known as data analytics. It is the process of analyzing and displaying data from hospitals or care services that have been stored in various business databases. Identifying the optimal hospital key performance indicators (KPIs) to compute, cleaning, scrubbing, and meta-tagging healthcare data to match the KPIs to measure, and then graphically displaying it using Tableau, Microsoft Power BI, Domo, or Qlik are all essential activities that must be done throughout a data analysis project.

Effective healthcare analytics provide organizational intelligence that helps to dig deep and through data and find the root of what the healthcare analytics initiative is telling in general. A healthcare analytics dashboard, unlike a static, excel-only report, can view real time, new, and meaningful data. Investing in analytics is therefore critical if you want to gain the benefits listed below.

Using underutilized and unstructured hospital data to create analytics assets that drive financial decision-making and outcomes.

More about the firm might be discovered through obtaining the appropriate type of healthcare findings, which would not have been possible otherwise. For example, let discharges be taken and split down into distinct groups.

The information provided by the healthcare dashboard will be used to perhaps reduce the number of readmissions at the hospital.

The costs of patients readmitted beyond 30 days reached $41.3 billion, according to a survey performed by the Agency for Healthcare Research and Quality. The cost of readmissions can be reduced by using the deep analytics system that was built into healthcare BI to keep a check on patient data, identify the patients with high-risk health risks, and monitor for the most frequent conditions that cause readmissions, such as behavioral or mental health issues.

Having real-time organizational healthcare analytics rather than a monthly manual Excel report

Instead of manually generating research for monthly reports, analytics should be created that can update in real time. The healthcare analytics dashboards should be refreshed with new data and let the app do the rest. This is more effective, less expensive, and more precise.

Having improved monitoring accuracy for hospital operations

Constant manual reporting leaves plenty of space for human error. Faulty data will result in higher costs and substantial rework to repair errors. A locked-down healthcare analytics dashboard, on the other hand, reduces the risk of human error, saving time, and resources.

Patients—be more informed of self-health

Some of the uses of predictive analytics include increased diagnostic precision, early diagnosis of a medical state in at-risk individuals utilizing genomics, and evidence-based treatment. Patients can be more conscious and confident of their own physical conditions in general thanks to the proliferation of wearables.

Physicians/Surgeons—Improve diagnostic accuracy

When a patient complains of chest pain, it can be difficult for a doctor to determine if the patient requires hospitalization. If the doctor is using a tried-and-true predictive diagnosis system that allows him to correctly input the patient's physical and clinical condition, the system will help the doctor make an informed decision. On the therapy side, a physician should track a patient's data (or electronic health record (EHRs)) over time and prescribe a treatment plan that is customized to the patient's unique condition. This evidence-based therapy lowers the likelihood of serious side effects.

Hospitals—Enhance patient care with low mortality rates

As illustrated in the case study above, predictive analytics may help hospitals and research center analyze the success of various operations and therapies in order to lower mortality and morbidity rates throughout the postoperative period.

Pharmaceutical companies—Can speed up the introduction of new, more effective pharmaceuticals to the market.

In the pharmaceutical sector, clinical trials performed for any new drugs are generally time-consuming, expensive, and involve resource-intensive procedures. Machine learning may be used to evaluate the permutations and combinations of already confirmed molecular components on a regular basis, making R&D operations in the pharmaceutical business more efficient.

This will assist in the development of new medications with a higher likelihood of success. Furthermore, predictive modeling may be utilized to examine the efficacy of novel medications more efficiently and cost-effectively. This will not only speed up the development of the medicine but also cut total healthcare expenses per patient.

Insurance Companies—Cut the cost of insurance

Predictive analytics models can be used by healthcare insurance providers to better forecast insurance costs for persons. The price of insurance is now decided by a person's age, current medical condition, and the "plan" they select. Now, thanks to developments in medical technology, genetic information, and other healthcare-related data can be freely shared.

Insurance companies can utilize this information to anticipate a person's future medical bills and make better-educated judgments regarding the insurance expenses connected with that person. This would result in a more accurate assessment of a person's insurance requirements, benefiting all parties in terms of the arrangements to be made.

Public Healthcare Professionals

The World Health Organization describes public health as both privately and publicly funded initiatives aimed at preventing illness, promoting health, and extending people's lives. Its operations aim to create healthier environments for individuals, with a focus on whole populations rather than individual patients or ailments.

Analytics can be used to forecast early detection of pandemics and flu outbreaks in this case. GoogleFlu was a project that exploited search patterns to forecast flu and dengue fever outbreaks. While the project is no longer publishing, the empirical data is still available for study.

Real-time monitoring of hospital statistics

An interactive platform for showing clinical data of some type is what a healthcare analytics platform is. It is a simple way to see data from different sources. However, data must be first standardized in a logical manner. Specific hospitals also have access to electronic health information, and the federal government has made a clinical trial and insurance reports more available as well.

As previously stated, each hospital maintains its own collection of EMRs for each of its patients. These records offer a lot of information on a patient's stay at the hospital. If an executive is dealing with several hospitals under one organization, the EMRs are almost definitely all the same. Starting with a standard format, compiling data is simpler. The state/federal information can be more difficult to obtain because it is gathered from a range of sources and may not be uniform.

Around the world, state governments and the federal government maintain a variety of medical data databases. The National Program of Cancer Registries (NPCR) of the Centers for Disease Control and Prevention (CDC) is one such example. This is a register that compiles data from local registries around the country on patient history, diagnosis, treatment, and status.

These databases are significant because they provide healthcare data analysts access to a larger pool of data. Depending on the reporting goals, it should be cross-referenced with government data to provide a better picture of patients. More significantly, there are more possibilities for finding the data needed to boost the development of healthcare data analytics.

Thanks to the diversity of public and private healthcare data accessible, analytics may graphically track KPIs and other BI variables across a wide range of healthcare processes. Healthcare analytics may reveal patient patterns, budget performance for certain departments, the rate of particular tests, and other details. This data may be used to identify how long patients have previously stayed at a hospital. Potential operational modifications in the admission-to-discharge pipeline might be found using this historical data.

The combination of data analytics and the hospital's admission-to-discharge channel has the possibility to benefit the hospital, the workforce, and the patients. The particulars can be drilled down to what goes into a patient's stay in the hospital by choosing the right KPIs for analytics. This data should be utilized to identify important issue areas that might affect the discharge process, such as the quality of care delivered and the number of employees on hand, and to address any delays. As a consequence, patients are happier and healthier, and expenses are lower due to shorter hospital stays.

The efficiency of data analytics dashboards is tied to the advantages of analytics in healthcare in many ways. If bad data is fed into the analytics, there will be bad dashboards. That is why it is critical to begin with the goal in mind before attempting to make any analytics. Before going through the data, business problem to be measured and addressed should be decided.

Data preparation and proper front-end KPI selection are essential for analytics dashboards. The success of healthcare analytics effort is determined by the chosen KPIs. KPIs in Healthcare—Healthcare KPIs are numbers or percentages that measure how well a hospital department, organization, process, or individual is achieving specified goals and objectives.

For a hospital or healthcare company, KPIs such as the *Average Patient Length of Stay* or the *Average Emergency Room Wait Time* are useful. The second step entails locating hospital operations data that is believed to be needed internally. All the datasets chosen must be used and manually analyzed to see if they are even relevant to the KPIs that are tracked.

Patient-related KPIs will not be included in every healthcare analytics dashboard. It is sometimes important to look further into operational issues that have a financial impact. The example of healthcare data analytics above provides for the visualization of current industry trends, such as real and goal revenue. The long-term effects of hospital income patterns may be assessed using data from the healthcare business.

6.2.2   *HEALTHCARE ANALYTICS OFFERS REAL-WORLD SOLUTIONS TO PATIENT CARE*

When implementing a data analytics strategy, the scope is crucial. Apply the healthcare metrics was assembled prior to rolling out analytics to real-world scenarios. One aspect to consider is patient data. By handling large data and visualizing it using analytics dashboards, it can create useful pictures of individual patients.

Reduced unnecessary tests

The use of big data in healthcare to eliminate duplication is a great example. Going to a doctor or a hospital and being compelled to take examinations that all appear to accomplish the same thing is something that no one enjoys. It is a waste of time and money, and it has no effect on the length of stay of the patient.

Furthermore, redundant testing can put patients' safety at risk. The research report, which was published by the journal of Health Affairs, proved that the tests have been redundantly taken by various doctors in hospitals. According to the same study, removing redundant tests will save the American healthcare industry $8 billion.

All of a patient's data, on the other hand, may be merged into a unified system utilizing healthcare analytics. The point is that integrating patient data makes accessing patient information via an EHR easier for clinicians. When a doctor is with a patient, they can see all the examinations and therapies that have already been performed. This facility saves money, eliminates the need for unnecessary testing, and allows doctors to provide better care to their patients.

Individualizing care

Individual patient data have become easier to access because of new analytics technologies. The ability to run personalized data through healthcare analytics comes with access to personalized data. One component of this, as anybody familiar with the healthcare sector knows, is talking to the patients about their medical data and making appropriate decisions, but healthcare analytics may also assist personalize another, more surprising, component: cost trends.

By communicating information in an easy-to-understand visualization, healthcare analytics becomes a useful tool for staff and patients to explain the cost of treatments. The same healthcare analytics dashboard will be used by both staff and patients, ensuring consistent communication.

Finally, this data must be reviewed in order to aid patients in making the best medical decisions possible. As a result, a more complete picture of the patient's wants and concerns develops. One of the benefits of analytics in healthcare is that it increases access to and exploration of individual patient data.

Individualized treatment regimens may one day replace today's larger, one-size-fits-all treatments, thanks to advances in healthcare analytics. The use of Big Data and analytics in healthcare may become more common as time goes on. Now, more than ever, healthcare analytics and dashboards must be pursued.

Technomedical Innovations

Over the last few decades, the healthcare industry has seen a number of significant technological advancements. EHRs, HIV combination medication treatment (HAART), minimally invasive surgery, needle-free injection technologies, MRI, genomics, and noninvasive diagnostics are just a few examples. These advances are significant because they improve, accelerate, and secure patient diagnosis and treatment.

Endoscopic surgery, also known as minimally invasive surgery, is one such scientific breakthrough. The way surgeries are conducted has changed dramatically as a result of this breakthrough. On Knowledge@ Wharton's list of the "Top 30 Innovations of the Last 30 Years," it was formerly placed on the tenth position. Traditional surgical techniques were much more invasive, risky, painful, and time-consuming.

They necessitated longer hospital stays and recovery periods after surgery. Today, thanks to technical advancements, patients can choose between robotic surgery and endoscopic (nonrobotic) surgery, all of which result in significantly shortened recovery times, pain, and scarring. This is good news for patients who want to get back to work as soon as possible.

Minimally Invasive Cardiac Surgery

For specific cardiac surgery procedures, many hospitals and cardiac care centers around the world are investigating the effectiveness of the newer—minimum invasive approach. In traditional cardiac valve repair/ replacement surgery, a patient's sternum was opened with an 18–20 cm vertical incision. The newest minimally invasive procedure includes making a considerably smaller incision beneath the right breast to access the heart (horizontal up to 7 cm or a keyhole).

Because the region of access (to the heart) is considerably narrower than in traditional surgery, the new cardiac treatments are more challenging for surgeons to execute. Despite this, the new procedure is thought to result in less bleeding, a lower risk of infection, quicker recovery times, and lower patient costs.

A client was comparing the efficacy of the new method of cardiac surgery to the traditional method of conducting the same procedure. The research helped them to enroll patients in one of two kinds of procedures based on a variety of physiological and health factors. They conducted the research over a 3-year period and kept track of their findings. The type of surgery conducted, as well as many relevant output metrics such as the length of hospital stay and pain levels experienced, were also gathered.

The data was collected and randomized for detailed analysis, and a power of 80% was selected. To understand the nature and importance of each output parameter, the analysts worked closely with the client. Based on the nature and type of data to be analyzed, the appropriate statistical tools and methods were chosen to use.

The studies were carried out with a 5% statistical significance level. The findings were thoroughly reviewed for statistical and clinical importance. The statistical test findings were cross-checked quantitatively and qualitatively with subject matter experts for completeness and accuracy to arrive at unambiguous results.

Each conclusion presented to the client was backed up by data. The findings will aid them in making an informed policy decision to adopt the newer method at their facility. A predictive model was constructed based on the findings of the initial study, which went beyond statistical analysis. Based on certain considerations, the predictive model can help the client in evaluating the best approach to take for a future patient. The cardiac procedure outcome at the hospital is expected to increase as a result of this.

The Analytics Emergency: Rapid Implementation of Real-Time Analytics during pandemic

The COVID-19 pandemic forced leveraging the technological advancements and analytics to make decisions, highlighting the importance of reliable real-time data and analytics. Albany Med's analytics platform had a 2-day lag for most of the data, despite providing plenty of patient data for direct patient care. The organization soon realized that easy access to COVID-19 analytics was critical and that having real-time data for decision-making was critical for leaders during the COVID-19 calamity.

Featured Outcomes

- In less than 16 h, integrated, real-time dashboards were implemented for monitoring all COVID-19 activities, including patient testing and activity, care requirements, and supply utilization.
- A 99.8% decrease in processing time, allowing the incident command to make educated choices with the crucial information they require. Thousands of patients had their COVID-19 testing done.
- Over 13,000 employees, helpers, and contractors have access to unified workforce and COVID-19 data.

Prior to the pandemic, Albany Med had access to large amounts of patient data; however, the data ingestion process resulted in a 2-day delay in visualizing the most up-to-date information through the data platform because data had to be moved from operational sources to source databases before being aggregated into the data platform. While this data flow architecture addressed the bulk of business objectives, it was insufficient for handling the exceptional COVID-19 outbreak's speedy reaction.

Workforce data for 13,000 personnel, volunteers, and consultants, in addition to patient data, were to be made useful. That information was scattered across different systems, so there was no one source of truth for tracking all the company's employees. Furthermore, there was no direct relationship between the HR system and the employee health system. Albany Med needs precise, real-time data in order to keep the staff safe and make choices in the mother.

Thousands of patients were treated at Albany Med for COVID-19 testing. Hundreds of faxes were received from the lab, and staff had to seek up individual patient phone numbers, as well as try to contact and document patient contact. COVID-19 activity, patient mobility, and outcomes were all tracked manually, which took a long time. Albany Med's executives required real-time data to make choices during the COVID-19 issue.

After the presentation of Albany Med's initial COVID-19 instances, a formal incident command was created. It was evident that the organization needed to include analytics into incident command decision-making and everyday operations under the new care delivery paradigm. Albany Med employed the Health Catalyst® Data Operating System (DOS™) platform and a sophisticated range of analytics apps to provide real-time data access and improve the organization's response to the pandemic. Albany Med data scientists developed strong COVID-19 incident command analytics in two rounds.

COVID-19 incident command analytics are developed in two stages, as shown below.

COVID-19 analytics development began with an emphasis on optimal patient care, providing answers to key issues such as:

- Who are the people who are going to be treated?
- Where are the patients physically placed in the hospital?
- Do they need to be tested?
- If they have been tested, are they waiting for the results?
- Do they require medical attention?
- What kind of care will be provided?

The second phase of the COVID-19 analytics focuses on keeping workers safe in the face of continually changing legislation, a shortage of personal protective equipment, fragmented systems, and insufficient tests. The main questions for phase 2 were:

- Who is employed by the organization?
- Which employees are ill?
- Which workers have healed completely and are ready to return to work?

Albany Med used DOS to increase the timeliness of data by using HL7 messages from multiple systems, ensuring that the organization could imagine changes as they happened and take action. To meet these demands, a real-time database was created, which included:

- Information about the patient.
- Visitor information (encounter type and clinical service, current location, and status changes).
- Orders for supplies (all lab order, electronic order indications, and details).
- The end outcome (raw lab results, assessment details, and other key-value information).

Albany Med created an epidemiology dashboard quickly, allowing for real-time monitoring of confirmed COVID-19 cases and individuals under investigation (PUI) as well as centralized results recording. It also designed a tool for infection preventionists and clinicians to keep track of contacts, exposures, and interventions.

Infection preventionists and physicians may now quickly and easily see all patients who have had a pending or positive COVID-19 test using a contact tracing and exposure monitoring tool. Albany Med was able to see patient mobility for all PUIs and patients who tested positive for COVID-19, including movement in and out of the clinic, total cases, ventilator use, intubation status, and ultimate disposition. These real-time visualizations have been put to excellent use for incident command and operations management, and they are still in use today.

Albany Med aggregated the multiple workforce systems into a single consolidated list, employed person-matching algorithms to match persons across systems, and linked COVID-19 test results to the relevant worker. The employee health department used the same people-matching technology to replace the first paper intake form with an electronic form. Workplaces, mask use, and symptoms were visualized and monitored by the company. It also kept track of which employees have been placed under quarantine by the state.

Worker contact tracking and decision support were used by the organization to guarantee that symptoms were reviewed at the right time to validate the necessity for continued quarantine or the capacity to return to work. Albany Med will look at how many healthcare workers have been exposed, as well as how many patients and staff have been exposed, how many nosocomial infections have occurred, and how many COVID-19 cases have been linked to healthcare workers.

To assist the safe resume of elective procedures, Albany Med built callback apps and data entry tools for preadmission testing, monitoring antibody testing, N95 fit testing, and the provision of suitable personal protective equipment. The organization put up a distinct contact tracking callback program for first responders who came for testing and treatment.

It also generated a report to track down businesses that brought in COVID-19 patients and notified them of the workplace exposure, allowing for follow-up. Albany Med has also aggregated data from a number of sources to track COVID-19 positive individuals in communal living facilities, such as nursing homes, treatment centers, and penal facilities.

Albany Medical Center's swift analytics reaction enabled the organization to respond rapidly to the COVID-19 epidemic by providing critical, real-time data, enhancing incident command effectiveness. The following are some of the outcomes:

- In less than 16 h, integrated dashboards were implemented to track all COVID-19 activities, including patient testing and activity, care requirements, and supply usage.
- A 99.8% reduction in processing time, giving incident command the critical information, they need to make informed decisions.
- Thousands of patients had their COVID-19 levels checked.
- Over 13,000 employees, volunteers, and contractors have access to integrated workforce and COVID-19 data.

6.3 BIG DATA ANALYTICS HEALTHCARE SYSTEMS

In this section, we elaborate on various verticals of big data analytics research techniques and methods in the healthcare system. Figure 6.1 depicts the same.

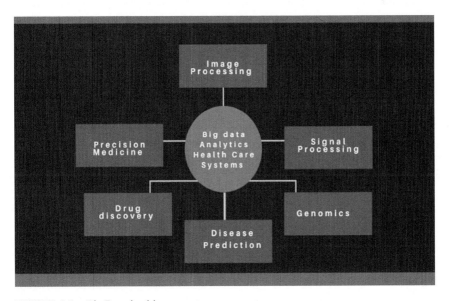

FIGURE 6.1   Big Data healthcare system.

6.3.1   IMAGE PROCESSING

Big data analytics for medical image processing helps us to identify the disease and in early diagnosing and treatment.

6.3.1.1   CLINICAL IMAGE ACQUISITION

Medical imaging is the collection of pictures obtained from numerous sources such as MRI, computed tomography (CT), X-ray, molecular imaging, ultrasound, and positron emission tomography (Gessner et al., 2013). Usually, the medical images are higher in resolution, dimensions of 2D, 3D, 4D, and size because it depicts the imaging of organs of the body. To handle higher end images with varied dimensions requires high-performance computing with advanced analytics. Various medical public datasets are available for research.

6.3.1.2   PREPROCESSING

This stage involves feature extraction, applying filtering to the medical images, cleaning of the same to clear noises in the captured image. This is the initial stage of analytics methods. The preprocessing is applied to various layers of the images. The data normalization can be applied at every slice of the image under study, for the entire image or a particular section.

The obtained medical images are prone to noises due to random noise, statistical noise, electronic noise, and round-off errors. In general, the images are denoised from the two perspectives one is spatial and the other one is transformed. The challenge here is to maintain the flat part of the image as same, keeping the boundaries of the image without a blur, maintaining the details of the texture, and avoiding newly included artifacts after preprocessing.

The conventional method to remove noise was spatial domain filtering. This method is applied directly to the image. This method is further split into linear and nonlinear. Linear methods were used initially but they lack in preserving some details of the images. Median filtering method is applied to reduce Gaussian noise but the disadvantage is that it produces blurry images when the noise is higher.

The transform-based filtering method is a kind of wavelet denoising that converts the image into low- and high-frequency coefficients by applying wavelet transformation. This is carried out by first estimating the noise variance followed by applying threshold on each part of the image.

Image data interpolation is the technique of confirming the physical spacing of the image is equally spaced. In order to maintain the same resolution, the center-specific or reconstruction of image is validated after preprocessing.

Image registration is to align the part of investigation with the reference image. The slice level image registration is carried out by slice level or image level due to the patient's movement during the scan. Image level registration is done for the training datasets in order to enhance the learning process of machine learning either classification or segmentation.

Some images pose a great challenge in analysis when it is soft contrast in nature, with lesser image intensity. Most of the images are degraded due to added artifacts when it is collected from different sources. The image has to be denoised for further analytics. Slice level registration is applied for processing 3D images.

Many research methods have been developed for standardizing the image which is carried out in two steps. First is shifting the intensity of the pixels to a range of positive numbers and then applying the probability function to remove the upper and the lower tail part of the histogram.

6.3.1.3 BIG DATA ANALYTICAL METHODS

The motive of medical imaging is to derive inferences after analyzing the big image datasets. Recently, many frameworks have been developed for experiencing enhanced data analytics. One of the models developed for investigate large medical datasets is MapReduce-based Hadoop architecture.

The advantage of the above-mentioned architecture is to:

- handle large-scale medical images,
- fine tuning the parameters for classification problems,
- texture classification—wavelet analysis, and
- medical image indexing.

For classification, well-known machine learning methods are applied. Even though the conventional machine learning algorithms are used for analytics, deep learning algorithms are evolved as the promising one, when we try to handle images of huge size and to solve various image classification or segmentation problems.

For example, in the case of diabetic retinopathy, the CNN architecture is used because of higher accuracy of the model that exceeds human recognition ability. This is mainly used to perceive fundus images, which in turn detects retinopathy and glaucoma. In addition to this vessel segmentation, pulmonary nodule screening, optic disc, and cup segmentation are the areas where CNN is applied.

Pathological analysis is the field of diagnosing cancer (Masoudi et al., 2021). The deep learning algorithms are used for breast, lung, and gastric cancer types. The analytic system is intended for early tumor screening and detection.

Cardiac and ventricular segmentation is done for analysis MRI images. This is usually done for measuring the thickness, volume, and ejection fraction. A recurrent full convolutional network that automates the learning process of 2D slice of the given image can further segment and detect heart disease. The added advantage of this method is reduced computational time with simpler segmentation process.

MapReduce-based optimal classifier is applied for diabetes mellitus classification, with gradient boosting tree algorithm. This method achieved higher accuracy when applied with diabetic datasets. The meta-heuristic algorithm based Cuckoo Grey wolf with naïve Bayes classifier based on map-reduce method is used for classification of medical images.

6.3.2 SIGNAL PROCESSING

Most of the medical devices emit signals that are recorded for diagnosing. The continuously generated data from these devices have to be monitored extensively for proper analysis which is not possible with the conventional methods.

Big data streaming health analytics is defined as capturing the streaming data from the devices that are in the continuous wave nature, then applying all the analytics techniques that are quantitative, cognitive, and predictive to draw the inferences and devising procedure for treatment with a lesser time.

Thus, big data analytics plays a vital role to combat all the difficulty in gathering and analyzing the signals degenerated by the sensor or other medical devices. There has been an increased attempt to utilize telemetry over a period of time for enhancing patient care and manage their health records efficiently. The key component of big data analytic systems is how we integrate statistical and dynamic elements of streaming data for analytics.

The big data streaming platforms are comprehensive in nature that can be implemented for this type of analytics. This system autoextracts the features, and then learning models are implemented with an incremental approach. The model's output might be descriptive, analytical, or prescriptive. This will serve as a warning or notification to the doctors.

Figure 6.2 depicts the flow of the procedure. The data ingestion stage collates the data from different sources, thus combining the static and dynamic data of the patients. The system model enables the auto-extract features. Finally, actionable insights are derived from the system as the outcome of analytics (Belle et al., 2015).

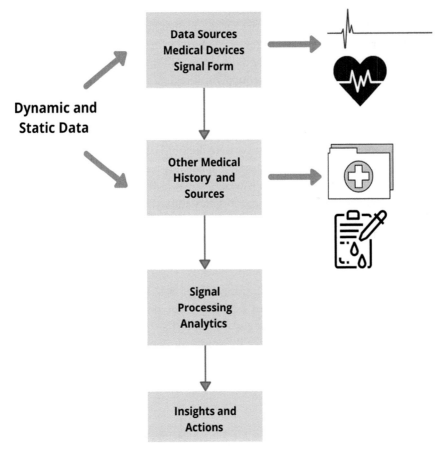

FIGURE 6.2 Signal processing analytics.

6.3.3   GENOMICS

Gene sequencing is an integral part of genomics. Genomic analysis is a vital part of clinical diagnosis after which treatment is given for the patient. The incorporation of physical and biochemical mechanisms together paves the way for enhanced analytics with valuable inferences and less cost-effective mechanisms.

The sequencing process is to map the genome references. The string processing machine learning algorithms are used for the same. Before DNA sequencing is carried out, protein pairwise alignment is done as a part of the analytics mechanism. The sequencing process is to map the genome references (Lander and Waterman, 1988). For this, machine learning methods for string processing are applied. Protein pairwise alignment is performed as part of the analytics procedure before DNA sequencing.

To combat the increase in the volume of genomics data, instead of text format storage of biological sequences like FASTQ, Binary Alignment Map format, and variant call formats are preferred. The various formats proposed in recent literature advocate the technical development with an increased complexity of cryoelectron microscopy and mass spectrometry.

When we look into the data growth in the field of genomics, data is escalated per nano seconds. The genomics analytics has considerably reduced the gene sequencing time from 8 weeks to 24 h, thus providing a quicker sequencing process. Minimum redundancy maximum relevance, maximal information coefficient, Monte Carlo feature selection, Cox-regression, SVM, Fireflies and ant colony, decision tree k-nearest-neighbor, and other preprocessing methods are employed in this procedure.

Human genome project paved the way for exploration of genomic sequence data. Recently, Hadoop-based genome projects are developed to accommodate the challenges existing right now. The bioinformatics domain makes extensive use of parallel and distributed architecture. The outline of the algorithms used for this purpose is presented in Figure 6.3.

6.3.4   DRUG DISCOVERY

Advancement in the domain of medical imaging and sequencing leads to abridged process of drug development. Drug discovery is the process that may require data acquisition (both structured and unstructured data) and analytics of data of huge volume in nature. Many research laboratories have been investing quality research efforts to develop analytics model in order to cut short the expense incurred during drug development.

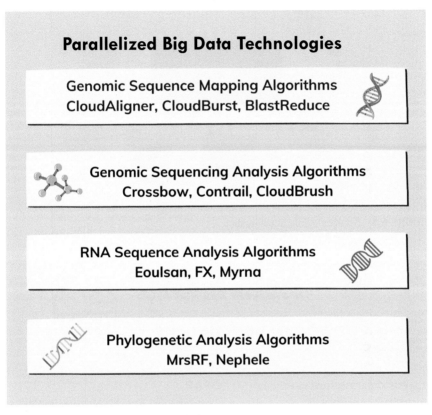

FIGURE 6.3 Genomics big data analytics.

The drug development process is usually slow and incurs huge investment because conducting the real-time experiment with toxic elements with lesser facilities. These challenges were addressed by the big data analytics model which stimulates the entire process with reduced cost with enhanced efficacy of medication.

The sources of drug discovery are depicted in Figure 6.4.

The various big data analytics methods employed for drug discovery and development are depicted in Figure 6.5.

FIGURE 6.4   Sources—drug development.

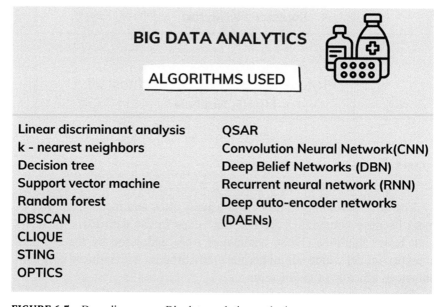

FIGURE 6.5   Drug discovery—Big data analytics methods.

6.3.5 DISEASE PREDICTION

In the present scenario, disease prediction is vital across the globe. The early recognition of disease is indeed crucial to prevent adverse effects of the disease. Big data analytic techniques mainly emphasize on the disease prediction and exploration of health status. But the conventional system lacks the capacity to assimilate the historical and streaming health data.

The Hadoop MapReduce-based artificial neural network classifier framework is suggested for disease prediction. The distributed, simple architecture solves disease prediction problems with enhanced accuracy prediction.

For example, map-reduce-based smart heart attack system predicts and prevents cardiovascular disease. Several machine learning methods are employed for disease prediction, which is presented in Figure 6.6.

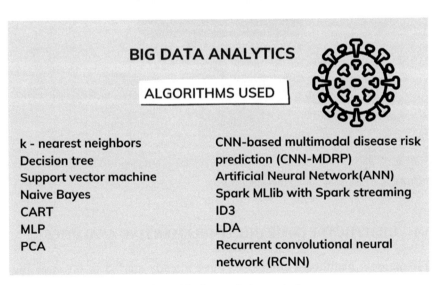

FIGURE 6.6 Disease prediction—Big data analytics methods.

6.3.6 PRECISION MEDICINE

Precision medicine is of the emerging research area, which emphasizes personalized treatment based on gene variability of an individual, the geographical location, lifestyle, and other personalized factors. This method paved the way for devising prediction strategies for a particular person in contrast with devising generalized strategies.

The sources for this type of analytics are depicted in Figure 6.7.

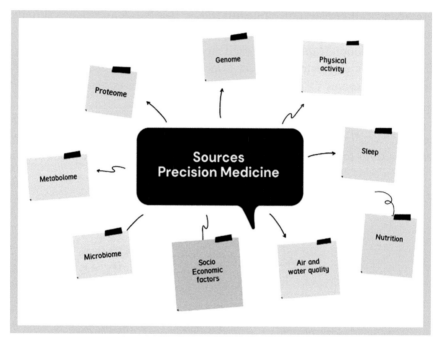

FIGURE 6.7   Sources—precision medicine.

6.4   HEALTHCARE COMPANIES IMPLEMENTING ANALYTICS

International healthcare companies have already started to leverage big data analytics.

The next sections outline a few.

Washington State Health Care Authority

The Washington State Health Care Authority is a factual example of data analytics positively affecting the healthcare sector. They, like many other healthcare institutions, see overuse and overcrowding in their emergency rooms, resulting in dwindling staff and increasing medical costs. The company wrangled patient data to classify repeat patients and distribute their data through several hospitals using an independently personalized data management initiative that included focused market intelligence.

Sisense—Union General Hospital

Sisense, headquartered in New York, provides market intelligence tools that help businesses plan, imagine, and derive useful information from their data more effectively. Sisense explains how it aided Union General Hospital, a nonprofit healthcare provider in Northern Georgia, in reducing data processing time from a day to five minutes in a case report. According to Sisense, the core team in charge of decision-making support manages around 20TB of data in excel files and relational database management systems. Examples of outcomes include digital patient charts, electronic medical records (EMRs), and hospital papers utilized in day-to-day activities.

The hospital's EMR system was inefficient, the CFO spent the majority of his time manually crunching data in Excel, and their IT services were limited. The hospital wanted a cost-effective market analytics approach that was also convenient to use. The hospital's head of decision-making service said if there is not a million-dollar budget and a workforce of 30 to fund healthcare analytics, there cannot be a frustration in nonprofit healthcare. So, Sisense helps to make improvements by remaining under the budget.

Owing to a lengthy logistical process, the hospital's 30-day readmission records could only be checked and published on a quarterly basis prior to introducing the Sisense BI scheme. The detail is now open to the case management committee on a monthly basis. Case managers with real-time monitoring skills spot ways to reduce readmissions as soon as they emerge.

Domo—Apria Healthcare

Domo, headquartered in Utah, was founded in 2011 and provides a BI tool to help companies handle their activities. In a case report, Domo describes its work with Apria Healthcare, a home healthcare firm with over 400 locations and over 1.2 million patients treated per year. Apria is an organization that deals in ventilation care and associated medical equipment including oxygen and sleep apnea treatment.

Prior to using Domo, Apria had trouble monitoring real-time data on sales results. This was partly due to siloed data collection and retrieval systems, which prevented a systematic and automated sales appraisal. Prior to Domo, the key difficulty, according to the Senior Director of BI and Data Analytics, was fragmented data, as well as difficulties with sales visibility. He believed the sales data were retrieved a month and a half after the acquisition, and it was not in the dimensions necessary to gain a complete view of the company.

According to the case report, Apria was able to obtain a consolidated view of its sales metrics after applying Domo. Real-time data monitored by Domo include cash balances, social media output data, and website traffic. Domo's data mining, according to Apria, aided the company's decision-making capabilities by providing visibility into where it got the majority of its sales and opportunities to expand market exposure.

Qlik—Children's Healthcare of Atlanta

Qlik Technologies, based in Philadelphia, PA, was founded in 1993 and provides BI and analytics tools to help businesses gain actionable information from their results. Qlik explains how it helped Children's Healthcare of Atlanta (CHOA) enhance its functionality in key areas such as logistics, financing, and research and development in a case study. One of the major providers of pediatric health services in the United States, Children's Healthcare of Atlanta is a nonprofit organization with three clinics and 16 neighborhood locations that care for over 500,000 patients each year.

It is important to note that, while certain data analytics needs are universal, each healthcare organization has its own set of criteria that are tailored to its patient demographics and success goals. Other BI platforms have been attempted by CHOA in the past, but they did not perform well for the company and caused friction between the IT and business divisions.

In the end, the Qlik platform proved to be a better match for CHOA's objectives. The Qlik platform resulted in a 65% reduction in reporting time, elimination of a 10-day waiting period for research-related requests, and new ideas for cost-cutting and quality-assurance initiatives.

Dundas—Family HealthCare Network

Dundas is a data analytics firm that was founded in 1992 and is based in Toronto, Canada. Dundas describes how it assisted a nonprofit community-based health group in meeting government targets in a case study. The Family HealthCare Network (FHCN) is a network of 16 federally accredited health centers that offer primary care to over 600,000 patients each year. FHCN wanted a BI approach to enhance data visualization and clarity across divisions.

The major goal of the healthcare organization was to improve its efficiency in the US Department of Health and Human Services' Uniform Data System monitoring. The data in this system are audited once a year to ensure that the health centers are following legislative and regulatory requirements.

Active referrals are an important aspect of FHCN's market activities; however, they started to outpace the number of consults earned, which could lead to other problems. Active referrals decreased by 20% after Dundas' BI platform was implemented, and report production time was cut in half.

While analytics can be used in all aspects of patient care, the focus in healthcare is on using the power of analytics to quickly assess the true value of technological developments for human benefit. Analytics enables us to make data-driven decisions about embracing and internalizing the latest technology developments that promise to improve our quality of life.

Flatiron Health

Flatiron is utilizing Big Data to assist doctors and scientists in the battle against cancer, one of the most important issues confronting doctors and scientists today. By automatically analyzing terabytes of data obtained during the diagnosis and treatment of cancer patients, the business expects that its oncology cloud will be able to extract insights from 96% of accessible patient data that has not yet been collected or processed.

It secured $130 million in funding last year, the majority of which came from Google, and it recently purchased Altos Solutions, a cloud-based medical records provider. The goal is to make their own data and knowledge accessible to as many healthcare providers as possible using their technology.

6.5   SOCIAL BIG DATA ANALYTICS

The storing in computer clouds of massive amounts of data from sources such as online browser data trails, social network interactions, sensor and surveillance data, and then looking for trends, fresh discoveries, and insights is known as Big Data. In less than a decade, Big Data has developed into a multibillion-dollar industry.

The huge quantity of data collected and gathered as a result of everyday activities and background facts, as well as the widespread use of social media in all sectors of life. The integration of social media into a variety of applications has brought up new opportunities for researchers. Because of the nature of social media interactions, a large amount of semantic data has been generated.

The constantly generated social media streams are devised as "Big Social Data." The application of big data in this domain is called social media analytics. Social media analytics directly or indirectly influence our life. On the flip side, this technology is intended for leveraging business activities also. Many companies have started using this platform for business in particular social media advertising has been increased.

We can anticipate this trend going to influence higher as this is identified as one of the factors for business development strategies. This mode of interaction is preferred because reaching out of people is made easier. Creating profile, generating brand messages are in trend now.

The reach of any business is to first identify the target group and then promote the business. Microtargeting is the most efficient way to augment the business activity, where a specific target group is recognized. Machine learning and Big Data technologies are used to do this. As a result, the integration of technology with media contacts will help in more competent corporate operations. The "Know Me" initiative from British Airways combines current loyalty data with information obtained from customers based on their internet behavior.

By merging these two sources of data, British Airways is able to create more tailored offers and respond to service gaps in ways that improve the flyer's experience. HiQ, a business that specializes in "people analytics," gathers public data, internal data from the organization, or a mixture of the two, as needed, to create predictive and practical models.

Thanks to clever algorithms, "machine learning" has provided more powerful insights than ever before (which is less time-consuming than human-powered legwork). The information is updated continuously and in real-time, and the reporting is customizable. Basically, it is the pinnacle of customized reports.

SumAll, the New York City-based startup focuses on assisting businesses in optimizing their social-marketing campaigns. The entire development is depicted in a single graph that includes data from 42 distinct platforms (obviously including the biggies like eBay and Facebook).

A Twitter alert system may be employed, which pings when new tweets are added, as well as a myriad of additional features that can be tailored to the user's needs. SumAll, on the other hand, is unusual in that the firm's nonprofit arm receives 10% of its ownership, allowing employees to feel good while doing well for the firm.

Netflix is now leveraging its vast database of data and analytics on global watching trends to produce and acquire programs that it knows will appeal to massive, ready-to-watch audiences. Netflix is now using its trove of data and analytics about international viewing habits to create and buy programming that it knows will be embraced by large, readymade audiences. After attracting millions of users with high-quality original programming, Netflix is now using its trove of data and analytics about international viewing habits to create and buy programming that it knows will be embraced by large, readymade audiences.

We are producing a product that actually appeals to the local preferences in each of the more than 50 countries where we operate, and it turns out that the local flavor is very universal, says Ted Sarandos, the company's chief content officer.

The name, Affectiva may not be familiar, but it may soon be part of our usage. They develop "emotional measurement technology," which is based on facial recognition and allows for the analysis of photos and videos to determine the mood and feelings of those depicted.

The technique may be used to measure the mood of individuals pictured engaging with a company's brand or service, or to gauge the mood of a political debate audience. Coca Cola and Unilever have both employed Affdex software for analysis, and the technology is projected to become more widely employed in the future in marketing and a range of other applications. Sentiment of the people is an important factor for business. The positive, negative, or neutral sentiments are vital for company's reputation, so analyzing these sentiments gathered by social media is really important. Leveraging big social data has many advantages, mentioned a few below.

Sources of data at the single point

The activities of the user are collected from various sources but they are done at a single point. For example, sign-ups, log entries, browser history, and number of clicks. The mobile applications are usually synced with the social account, also enabling the companies to gather the information at a single point.

Real-time communication

User's activities are tapped immediately and processed. The real interaction with people and feedback is done with lesser time through forum, follows, comments, tags, and so on. This is the best podium for real-time data analysis. The incredible changes in the technology accentuate us to leverage these benefits into revenue.

Target clients

Client identification is done through the social media platform. Machine learning with big data analytics gives the possibility of analyzing user preferences, their personal data, and social activities into some meaningful insights.

Prediction

Predictive analytics has been improved in a greater way in recent times, thus providing us on-time solutions in a lesser time. Big social data is derived from the functionalities of different domains namely big data science, data analytics, social computing, and computational social science. This is represented in Figure 6.8.

To handle and evaluate social media data, big data science delivers improved tools and approaches. Data analytic techniques are broader in terms of descriptive, exploratory, predictive, and prescriptive in nature. Computational science is the approach of applying a model to the data and drawing the inferences. It is an intersection of social and computational science.

It is the integrated approach collating social, behavioral, and cognitive theory together. The challenge here is to unravel the complex social data into meaningful ones. The core foundations of social computing integrate psychology, societal influences, social network analysis, and geographical location.

The sources of data are from various social media platforms. The data gathered are further moved into the data cleaning and preprocessing stage. Computational intelligence techniques are used after the data is ready for analysis. The final stage of analytics is to use visualization tools to show the results in a more appealing and intelligible manner. The overall workflow of social media analytics is depicted in Figure 6.9.

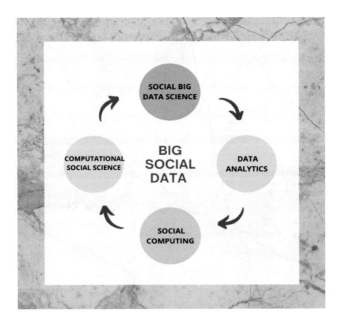

FIGURE 6.8 Big social data.

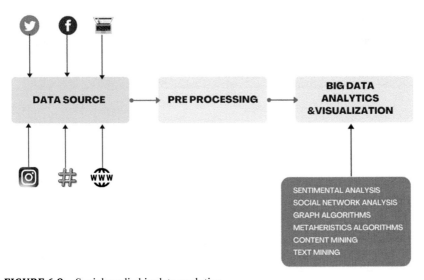

FIGURE 6.9 Social media big data analytics.

6.5.1   ALGORITHMS

This section explains the fundamental methodologies and techniques used in network analytics, community discovery, text analysis, information diffusion, and information fusion, which are all areas where information from social-based sources is now analyzed and processed.

6.5.1.1   SENTIMENT ANALYTICS

Sentiment analysis is a branch of social mining, where natural language processing technique is employed for classifying human emotions. Understanding of sentiments is important for business, medical, sociopsychological domains.

The various types of sentimental analysis are given below:

- Part-by-part analysis (fine-grained).
- Aspect-based analysis.
- Emotion detection analysis.
- Intent detection analysis.

The various divisions of sentiments are presented in Figure 6.10.

SENTIMENT TYPES		
POLARITY	**EMOTIONS**	**INTENTIONS**
Positive	Angry	Interested or Not
Negative	Happy	Urgent or Not
Neutral	Sad	Liked or Not

FIGURE 6.10   Sentiment types.

Fine—Grained Method

This method is based on feedback analysis, where the polarity of each input is mapped and results are derived. But this method is quite comprehensive and cost-intensive. The polarity types (positive/very positive/negative/ neutral) are identified and awarded with the weightage (Figure 6.11) five points/one point). The cumulative points are taken for analysis.

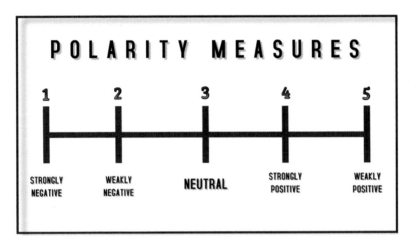

FIGURE 6.11 Polarity scores.

Emotion detection

This type of analysis is intended for detecting emotions (happiness, sadness, and anger). The lexical analysis technique is employed to map the words with emotions. But the challenge in this type of analysis is that sometimes the lexicon might wrongly predict when the user represents the emotions in different ways (Like killing it).

Aspect-based analysis

This type of analytics is used to find the aspect of emotion. For example, the product's sentiment is identified with particular features or character-istics of the product like lifetime, battery life, usage, and so on.

Intent-based Analysis

Intention analysis paves the way for better understanding and intention of the human. The intention of the customer's tastes to purchase a product or not is well analyzed. The inferences will help in targeting them by establishing patterns amongst the customer for product marketing or advertisement.

The algorithms are further classified into rule-based, feature-based, and hybrid methods.

Rule-based Algorithms

The system creates a rule, and the analysis is carried out based on that rule. The opinions, aspects, and polarity are considered for evaluation. The rules are crafted using NLP approaches. The various processes of NLP methods are presented in Figure 6.12 and discussed below.

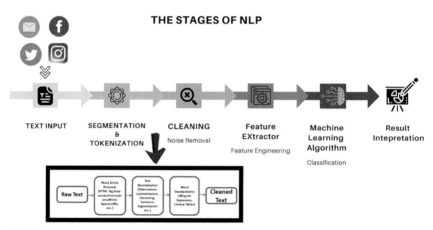

FIGURE 6.12   NLP algorithm.

During the training phase, the NLP tools convert the input into tag forms, which are further analyzed to establish the relation between the tags with the corresponding input. Statistical methods are used to form a word bank that contains the feature with semantics (Figure 6.14).

The various processes of NLP methods are discussed below.

Tokenization

The given strings are divided into a number of tokens. The delimiters are used to split the given sentence into words.

Tagging

From the given input sentence, the parts of the speeches are formed into the separate list to establish meaning.

Parsing

Dependency parser checks how well the ways are connected to each other (Figure 6.13). This establishes the grammatical structure of the sentence. Constituency method checks on phase grammar structure.

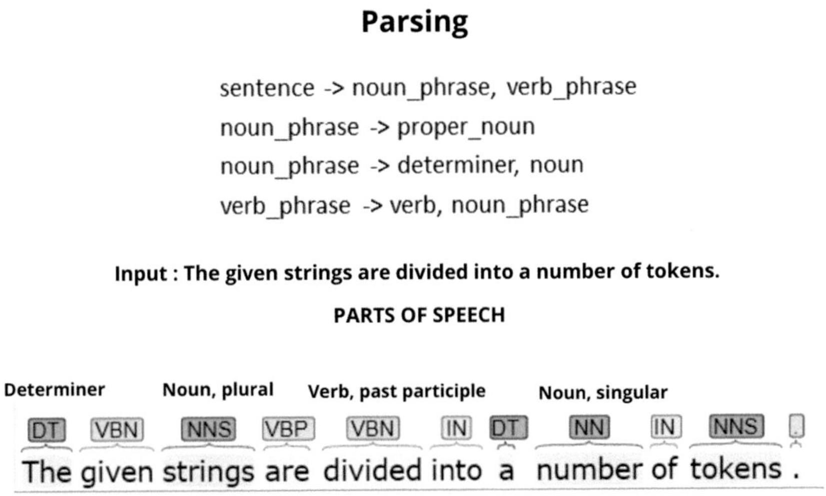

Parsing

sentence -> noun_phrase, verb_phrase

noun_phrase -> proper_noun

noun_phrase -> determiner, noun

verb_phrase -> verb, noun_phrase

Input : The given strings are divided into a number of tokens.

PARTS OF SPEECH

Determiner	Noun, plural	Verb, past participle	Noun, singular

DT	VBN	NNS	VBP	VBN	IN	DT	NN	IN	NNS	.

The given strings are divided into a number of tokens .

FIGURE 6.13 Parsing.

Lemmatization and Stemming

NLP employs lemmatization and stemming to convert words into their root form. The given words are represented in different forms and are converted to the root form called lemma.

For example:

Words	Stem
writing	write
achievement	achiev
studies	studi
copies	copi
believes	believ

Words	Lemma
believes	belief
copies	copy

From the example, we can understand difference between the lemma and stem representation. Lemma representation gives more meaningful form of a word (Lemma: believes → belief, stemming: believes → believ).

Feature-based Algorithms

This method is quite the opposite of the rule-based approach. Rule-based classification is completely avoided here. The classifiers are used to segregate the given into a number of output classifications. During the training phase, the classifier algorithm tries to correlate the input with the output feature. The feature vectors are generated by the feature extractor, which are further analyzed into predictable tags (positive, negative, or neutral). The overview of the various algorithms is shown in Figure 6.14.

NLP ALGORITHMS

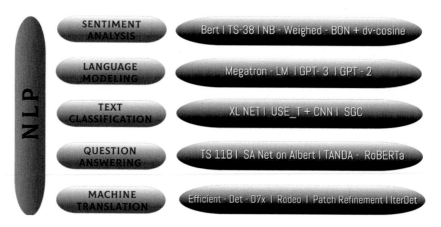

FIGURE 6.14   NLP algorithms.

Hybrid methods

This is the combination of both rule- and feature-based concept.

6.5.1.2   SOCIAL NETWORK ANALYSIS

At this time, the lives of everyone on the planet are intertwined in some manner. Network technology facilitates the intercommunication, sharing of information, emotions, and so on. Every second, across the globe social data is escalated at a higher speed. Network has become an important tool for information extraction and intelligence generation.

Social network analysis is primarily involved in identifying the pattern of people connected in a network. The connection is technically represented by graph or interconnected networks. The basic elements of graphs along are shown in Figure 6.15.

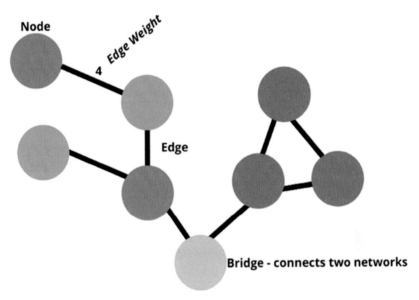

FIGURE 6.15   Social network graph.

In Figure 6.15, the graphs represent nodes or vertices represented as a small circle. The edge connects two vertices. We have to correlate this graph structure with the social network as:

User—Node
Edge—Relationship

Metrics used for graph analysis are depicted in Figure 6.16.

Various algorithms have been developed for graph analysis, which are presented in Figure 6.17.

Metrics	Definition
Degree	A node's degree is the number of edges the node has
Closeness	Closeness measures how well connected a node is to every other node in the network.
Betweenness	Betweenness measures the importance of a node's connections in allowing nodes to reach other nodes
Network Size	Network size is the number of nodes in the network.
Network Density	Network density is the number of edges divided by the total possible edges.
Length	Length is the number of edges between the starting and ending nodes
Distance	Distance is the number of edges or hops between the starting and ending nodes following the shortest path

FIGURE 6.16 Graph analysis—metrics.

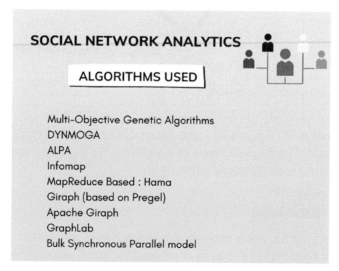

FIGURE 6.17 Graph analysis algorithms.

6.6 BIG DATA IN BUSINESS

Analyzing enormous volumes of data is what big data analytics is all about. This is done to reveal hidden trends and correlations, as well as provide insights into how to make sound business decisions. Essentially, businesses have acknowledged the importance of transitioning from a knowing to a learning organization. Businesses are adopting the power of data and technology in order to become more impartial and data-driven.

It is likely that not all the company's documents have been digitized. But then, big data encompasses all the company's meaningful, transactional, structured, and unstructured data. Big data analytics is important because mostly the unstructured data, which accounts for nearly 80% of all data, can help companies uncover hidden opportunities, better understand their clients, and optimize business processes.

When it comes to storing big volumes of data, tools like Hadoop and online analytics software may save a lot of money while also allowing you to establish a variety of cost-effective business models. Decision-making is both quicker and easier. Businesses can interpret information in real-time and make choices due to Hadoop's quick computing capacity and inbuilt memory analytics, as well as the extensibility to examine new sources of data.

The ability to quantify consumer preferences and have peace of mind through analytics opens the door to enable consumers to communicate their needs. Several firms, according to Davenport, are employing big data analytics to produce new commodities to meet customer desires.

Big data is a concept that has been around for a long time. Businesses employed analytics to get insights and detect patterns from data they collected decades before the phrase "big data" was created. This included manually inputting and comparing figures into a spreadsheet. Big data is analyzed using advanced computing tools. Companies are encouraged to reduce the time it takes to analyze data in order to make faster choices as a result of this.

In general, modern big data analytics systems enable quick and efficient analytical processes. Businesses gain a competitive advantage by being able to work harder and with greater agility. Meanwhile, organizations can save money by using big data analytics software. Big data analytics has become a priority for businesses. Consider a company that relies on rapid and agile decision-making to stay competitive.

To accommodate the wide scope of big data, businesses must be able to execute a structured evolutionary process. To do so, firms must first gather internal data in order to gain clear, actionable insights. The value of a company's comprehensive analytic process is higher. A strong big data analytics system should aid in detecting possible hazards or areas of vulnerability.

For a growing company, big data analytics is a critical investment. Businesses can gain a competitive advantage by introducing big data analytics thereby lower operating costs, and increase customer retention. Businesses should take advantage of a variety of customer data sources. Data is becoming more easily accessible to all companies as technical advances continue. It is reasonable to assume that firms currently have data at their disposal in terms of technology. Individual companies are in charge of ensuring that the requisite data analysis systems are in place to handle large volumes of data.

It is impossible to resist all of the big data hoopla. If companies have actionable data, they can better sell to customers, create and make things to fit specific demands, increase revenue, optimize processes, anticipate more correctly, and even better manage inventories to keep costs in check. Small enterprises need the same tools that bigger firms do to compete in today's market. Small firms, of course, lack the resources of a large company, such as data scientists, analysts, and researchers.

Your small business may use a variety of methods to gather, understand, and make sense of the data it already has, as well as get more insights to help level the playing field. The following are some common examples of big data analytics and how they can affect your daily business activities.

For years, big data analytics has been a well-known notion in digital transformation, but many firms also fail to capitalize on big data and its business implications. A strong IT infrastructure is critical for enhancing productivity while also reducing costs and maintaining security.

The more data your company will interpret, the better your defense will be. Big data analysis should be used by our security solution to get a clearer picture of what is "normal" in your company—who signs on when, who has access to what information, and how data is managed. IT would highlight and monitor any deviation from anticipated patterns in the company network, making it considerably more difficult for hackers to attack organizations that employ big data analytics. This is a common threat hunting tactic used in cyber security solutions.

Beginning in 2017 and continuing into 2018, Deloitte identified the use of people analytics as a major trend. Analytics is used in human resources for a variety of reasons, including:

- Sorting resumes and cover letters throughout the recruiting process.
- Analyzing video interviews to evaluate a candidate's personality.
- Recognizing behavioral trends in workers and departments.
- Gathering information on staff energy, well-being, and pain points.
- Identifying areas of payroll leakage or inadequate hourly time management.
- Choosing workers based on their quality and reliability.
- Tracking the real-time benefits of training and employee coaching.
    When analytics first appeared in marketing, companies began to learn how to best tempt clients to respond to their sales efforts through value propositions and calls-to-action. Since then, marketing analytics has shown to be valuable for a variety of reasons.
- Using big data analytics assist organizations in acquiring a better knowledge of market segmentation and prospective consumers.
- Gain a better grasp of customer preferences and behavior.
- Experiment with new products and more effective marketing strategies.
- Identify the most effective methods for improving the user experience.
- Make A/B testing more convenient.
- Assist in the development of pricing strategies.
    Businesses benefit from big data because it allows them to:
- Quickly identify the root causes of issues, defects, and failures.
- Create coupons based on a customer's purchase behavior in real time at the point of sale.
- Recalculate an entire risk portfolio quickly.
- Detect fraudulent or malicious cyberactivity before it has the worst possible consequences.
- Inspire the rest of your data and analytics strategy

6.6.1 ANALYTICS METHODS

The data injection phase starts accumulating the business data from various resources like people, product, promotion, price, and place (Fan, 2015). The sources of business data are shown in Figure 6.18.

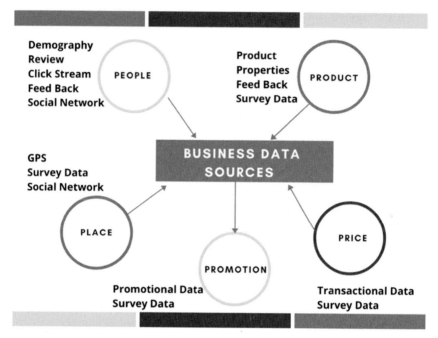

FIGURE 6.18 Business data sources.

Various methods of data collection are done through primary and secondary methods. This is done via both online and offline modes as well. The data from the board five categories are collected for further data analytics. Both static and dynamic data are collated before the analytics.

The different analytic algorithms incorporated are discussed below.

Customer Segmentation

Successful business strategy depends on the target and segment customers and devise business strategy accordingly. Customers with similar preferences and likes are categorized by big data techniques. Clustering algorithms are employed for segmentation problems. Various clustering algorithms have been developed in the field of machine learning.

The foundational theory of clustering is to identify the pattern amongst the given customer details and try to establish the pattern then group the similar patterns together. Clustering helps to enhance marketing with the specific target groups to address them effectively with a specific requirement. Companies can strategize the target marketing techniques to elevate the business which in turn secures the maximum profit. This will also help in completely evading risk in business.

Figure 6.19 depicts how clustering algorithms are used for segmentation problems.

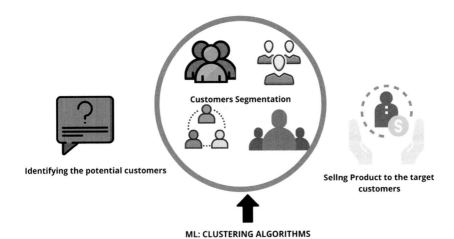

FIGURE 6.19 Customer segmentation.

Ontology Reputation Model

The feedback and product scores are used to create the reputation model. Both qualitative and quantitative data are collated by analysis. Semantics reviews also taken for consideration. This model extracts the features from the product's opinion and also counts the frequency of both positive and negative sentiments.

Aggregation algorithms are used to merge the customers' ratings. The reputation of the model is projected based on the following perspectives.

Frequency calculation (positive, negative/neutral)

Impact (Weightage is given for the feature based on opinion)

Product reputation

To validate the feature and produce a reputation score, fuzzy algorithms are frequently utilized. An ontology-based method is used for extracting semantics from textual input and defining the data domain. In another approach, ontologies are used to capture the tweets by providing a reputation model. Both entity and domain level data are analyzed.

The autogenerated ontology recommendations are useful for business development. But to foresee the changes in data escalation leads us in search of an enhanced and optimized big data approach. This is still under research and development.

Recommender System and Promotional Marketing

Promotional marketing is indeed necessary for any business. In this competitive world, investments are made higher for these kinds of activities to attract customers. The machine learning model projects the recommendations based on analytics.

The major classification of recommendation algorithms is given below:

- Content-based filtering
- Collaborative filtering
- Hybrid systems

Content-based method

This method learns the complete pattern of user preferences, their likes. A complete profile of the user is created to provide the knowledge about the specific customer (Figure 6.20).

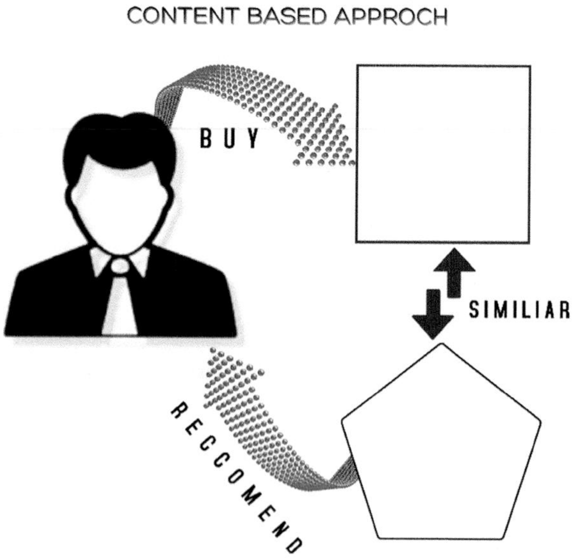

FIGURE 6.20 Content-based approach.

Collaborative Approach

This approach focuses upon finding the similarity or pattern amongst the customers (Figure 6.21). This is completely based on the user's opinion instead of system generated one. This is also called as a lookalike filtering method. Filtering is either user to user filtering or item-to-item based. User-to-user approach recommends based on the profile of the customer. The disadvantage of this method is to scale for large datasets. The item to item will provide recommendation on a case-by-case basis.

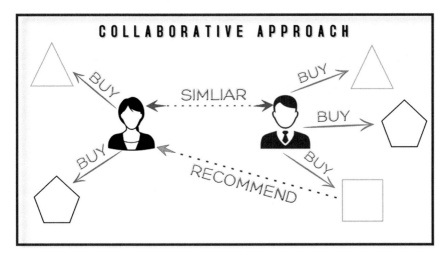

FIGURE 6.21   Collaborative-based approach.

Hybrid System

This method employs the functionalities of both content and collaborative approach.

Pricing

Fixing up the optimal price is needed for any business analysis. Companies should be able to identify the potential customers and the price that they are willing to pay toward the product. For this, demand analysis is undertaken by analysis of sales data of products and cost of the equivalent product in the market. This type of pattern analysis help in determining and establishing the product's best pricing.

The representative big data model for pricing strategy is presented in Figure 6.22.

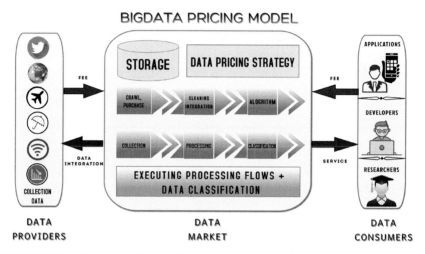

FIGURE 6.22 Big data pricing model.

In the above model, the first stage of data gathering combines data from all the sources. The classification algorithm is used to create marketing price plans. The price mix of a certain product is determined by the results of the analytics. The performance of a machine learning model based on ANNs in pricing prediction has increased. CNNs, in particular, are utilized for training and classification.

The key advantages of having data analytics are presented below.

Risk Management

Better risk management procedures are required in today's unprecedented times and highly risky business climate. In general, a risk management strategy is a crucial investment for any company, regardless of industry. If a business wants to be successful, it must be able to anticipate and minimize risks before they occur. According to business gurus, enterprise risk management comprises considerably more than ensuring that your organization is adequately covered.

Risk management must be approached strategically in every industry. While there are many factors that would contribute to this, big data is one of the most significant. It will also enable businesses to calculate and model the risks that their operations may face. Because of the usage of big data, companies may rely on predictive analytics to have informed risk foresight. If you have this knowledge, risk mitigation techniques will be easy to implement.

Big data analytics has made a significant contribution to the growth of risk management solutions thus far. The tools available enable companies to calculate and model the risks they face on a daily basis. Big data analytics has a lot of potential for improving the quality of risk management models because of the expanding availability and diversity of statistics. As a consequence, a company's risk mitigation techniques and strategic decisions should improve.

Know Customers Better

Big data enable businesses to better understand their customers. From website visits to social media interactions, big data may tell a lot about your customers. It also enables the company to build buyer personas and customer profiles. You can enhance your goods and services by better understanding them. It enables you to provide the highest level of customer satisfaction possible. This will also aid in the development of customer loyalty.

Identify Competition

You may also discover more about your competitors through big data. It will, for example, include details on their pricing models as well as how customers understand them. You will also hear about your rivals' customer attitudes. It can also assist you in determining their online performance, such as their social media engagement.

Individualize Marketing

The marketing strategy of the firm will determine whether it succeeds or fails. Personalization is one of the most significant aspects of its success. Big data can be valuable in this field as well. Marketing campaigns should be produced targeting a particular niche or segment because it allows us to understand the clients. It will provide information that allows advertisers to produce materials that convert well.

Look for Trends

Businesses should also use big data to detect trends, which can be helpful in product research and growth. Big data will give management the knowledge they need to understand how trends will change over time, allowing the firm to plan for these shifts in customer behaviors and purchase patterns.

Handle Human Resources

Big data should assist you in developing a solid human resource management strategy. Big data would aid in the recruiting, training, and growth of the best individuals. You will get the information you need to understand and improve your workforce, from employee satisfaction to productivity data. In the business world, big data has a plethora of applications. To make proper business decisions, businesses must find hidden trends, correlations, and insights from unstructured big data. From risk management to human resource management, big data may be utilized to improve business operations in a number of ways. As a result, ensure that suitable infrastructure for handling large data is in place.

Customer Acquisition and Retention

Any company relies on its most valuable asset: its customers. There is no company that can call itself successful without first building a strong customer base. Even with a customer base, though, a company cannot afford to ignore the fierce competition it faces. It is all too simple for a corporation to start offering low-quality items if it takes too long to find out what its customers want. Finally, you will lose consumers, which will have a detrimental influence on the overall performance of your company.

Businesses will use big data to observe numerous customer-related patterns and trends. In order to evoke loyalty, it is necessary to watch client behavior. Theoretically, the more data a business generates, the more patterns and trends it can discover. In today's business world and technological age, a company can easily collect all of the customer data it requires.

This means that the modern-day client is relatively simple to comprehend. To summarize, all that is necessary is a big data analytics approach to optimize the data at your disposal. A company would be able to derive essential behavioral insights that it needs to act on to keep its customer base if it has a suitable customer data analytics mechanism in place. Understanding customer insights will enable your company to provide exactly what your customers want. This is the first and most important step in achieving high customer retention.

Resolve Advertisers Issue and Offer Marketing Comprehensions

Big data analytics should aid in the transformation of all business processes. This involves the ability to meet customer expectations, modifying the company's product line, and, of course, guaranteeing the effectiveness of marketing campaigns. Let us face it, we are dealing with the facts. Businesses have squandered millions of dollars in ineffective ads. What is causing this? It is very likely that they missed the study stage.

After years of cautious optimism, the marketing and advertising technology industry can now fully accept big data (Medal, 2017). A more complicated analysis can be done by the advertisement and advertising industry. This entails tracking online activity, monitoring point-of-sale transactions, and detecting dynamic changes in customer patterns on the fly. Collecting and analyzing customer data is required to gain insights into their behavior. This is accomplished using a method similar to that used by marketers and advertisers, as shown. As a result, more concentrated and tailored advertisements should be run.

Businesses can save money and increase productivity by using a more focused and customized advertisement. This is because they use the right brands to target high-potential customers. Advertisers benefit from big data analytics because it allows them to better understand their customers' buying habits. We cannot ignore the massive issue of ad fraud. Organizations should identify their target clients with the help of predictive analytics. As a result, corporations would have a more reasonable and successful reach while preventing the massive losses caused by ad fraud.

Netflix is a great example of a huge brand that uses big data analytics to target advertisements. Netflix, which has over 100 million users, collects a lot of data, which helps it achieve its position. It makes use of this information to provide movie and television program suggestions to each subscriber. This is mostly done with the help of your past search and viewing history. This information is utilized to determine what the subscriber's primary interests are. Take a peek at how Netflix obtains massive data in the graphic below.

Product Development and Innovations Driver

Another important advantage of big data is its capacity to help firms with product creation and redevelopment. Big data has essentially become a means of generating new revenue streams by encouraging product innovation and improvement. Before designing new product and lines or redesigning existing products, organizations start by correcting as much data as much as it is theoretically feasible.

Every design process should start with determining what exactly fits the customers' needs. An organization should research customer needs through a variety of channels. The company would then use big data analytics to determine the best way to capitalize on that need.

"Those were the days when you could trust your instincts" (Rampton, 2017). A lot of data is needed to enhance the quality and streamline the manufacturing process. If a company tries to compete in the 21st century, it can no longer rely on gut instinct. This means that many businesses will need to devise ways to keep track of their products, rivals, and customer input.

After the data has been gathered, an analysis is carried out to ensure that logical reasoning is used before a plan of action is formulated. Fortunately, when it comes to compiling and using big data, product manufacturers of all sizes have a distinct advantage. As a result, these businesses can easily expand their product line by developing innovative products.

Amazon fresh and whole foods are two well-known names. This is an excellent example of how big data will aid in product development and innovation. To break into a global market, Amazon uses big data analytics. Amazon now has the necessary expertise in data-driven logistics to encourage the creation and achievement of higher value. Amazon whole foods should understand how consumers buy groceries and how vendors communicate with the grocer by focusing on big data analytics. When it is time to make more improvements, this information comes in handy.

Supply Chain Management

Big data provides higher precision, consistency, and insights to supplier networks. Suppliers can gain contextual intelligence in their supply chains by using big data analytics. Essentially, vendors can avoid the restrictions they faced previously by using big data analytics. This was accomplished using classical business management and supply chain management systems.

Because these legacy applications did not use big data analytics, manufacturers lost a lot of money and were more likely to make mistakes. Due to modern methodologies based on big data, suppliers, on the other hand, would benefit from increased degrees of contextual information, which is crucial for supply chain success.

More sophisticated supplier networks are possible with modern supply chain systems based on big data. To reach contextual intelligence, they are based on knowledge sharing and high-level cooperation. It is also worth noting that supply chain executives regard big data analytics as a disrupting technology. This is based on the belief that it would lay the groundwork for effective change management of companies.

PepsiCo is a consumer packaged products firm that uses big data to manage its supply chain. The company is committed to ensuring that appropriate volumes and varieties of items are refreshed on merchant shelves. Clients supply data to the business detailing their warehouse and point-of-sale inventories, which are utilized to reconcile and anticipate production and shipment requirements. This guarantees that suppliers have the correct items, in the correct amounts, and at the correct time.

Big Data Solutions in Business

Alteryx

Complex BI analysis does not have to be difficult. Alteryx uses powerful data mining and analytics techniques to show information in a clear and understandable manner. Alteryx mixes internal data from your company with publicly accessible data to assist you in making better business decisions. From the dashboard, you can make graphs, narratives, and interactive graphics using these perspectives. It also has collaboration tools that allow for group discussions.

In addition to corporate data, Alteryx may give department-specific data, such as marketing, sales, operations, and customer analytics. The platform covers areas such as retail, food and beverage, media and entertainment, financial services, manufacturing, consumer packaged goods, healthcare, and pharmaceuticals. Contact the company for pricing details.

Kissmetrics

Kissmetrics gives you the power to analyze, segment, and engage your clients based on their actions. Kissmetrics allows you to build, manage, and automate the distribution of single-shot and continuing email campaigns based on consumer behavior. Beyond opens and clicks, the platform assesses the effectiveness of campaigns. Kissmetrics for E-commerce, a new product from the company, is intended to boost your Facebook and Instagram ROI, lower cart abandonment rates, and encourage repeat purchases.

Web-based training and instructional tools, such as marketing webinars, how-to manuals, articles, and infographics, are available to Kissmetrics users to assist enhance their marketing initiatives. As part of the onboarding process, a personal customer success representative is assigned for the first 60 days, as well as strategic assistance to help you get the most out of the platform. Plans start at $300 per month.

InsightSquared

With InsightSquared, you would not waste time mining your own data and laboriously analyzing it with one spreadsheet after another. Instead, InsightSquared automatically collects data and extracts actionable information from major business systems, such as Salesforce, QuickBooks, Google Analytics, and Zendesk.

Using data from CRM software, InsightSquared, for example, will deliver a plethora of sales insight, including sales and pipeline projections, lead creation and tracking, profitability analysis, and activity monitoring. It will also help you detect patterns, strengths, and weaknesses, as well as your sales team's triumphs and failures.

InsightSquared also provides marketing, financial, employee, and support analytics, as well as bespoke reporting that allows users to slice and analyze data from any source in any way they choose. InsightSquared offers a free trial and service options that are flexible and expandable.

Google Analytics

To start collecting data, there is no need for any fancy, expensive software. It can begin with a resource that is already possessed: the website. Google Analytics, Google's free digital analytics product, allows small companies to analyze website data from all touchpoints in one location.

Google analytics can be used to extract long-term data to uncover patterns and other useful information so informed decisions can be made based on data. Monitoring and assessing visitor behavior, such as where traffic originates from, how viewers engage, and how long visitors remain on your website, may help you make better decisions to achieve your website's or online store's goals (known as your bounce rate).

Traffic on social media may be analyzed, allowing you to fine-tune your social media marketing strategies based on what works and what does not. Mobile visitors will be studied in order to extract information about consumers who visit the site on their mobile devices so that the mobile experience may be improved.

Qualtrics

Qualtrics enables to conduct a wide range of studies and surveys in order to gather valuable data for data-driven decisions. Qualtrics recently unveiled Qualtrics Experience Management (Qualtrics XM), a collection of four tools that help you enhance and sustain the experiences your organization provides to all stakeholders, including clients, employees, prospects, users, partners, suppliers, people, students, and investors.

Qualtrics XM enables you measure, prioritize, and optimize the customer, employee, brand, and product experiences you provide across the four basic company experiences. Qualtrics also provides real-time data, survey software, ad testing, and market research programs. Employee polls, exit interviews, and reviews will all be handled by the firm.

These big data analytics examples, at their core, demonstrate how data analysis can help companies become more cost-effective, productive, and competitive in their respective markets. Small- and medium-sized businesses are the most likely to use data and analytics tools to obtain a competitive advantage. Analytics and big data, when appropriately coupled, may provide helpful corporate knowledge about processes as well as new prospects. Data analytics in IT and cybersecurity help firms stay ahead of threats to secure customer, employee, and corporate data, which is extremely critical in today's cybersecurity environment.

6.7 EDUCATIONAL DATA ANALYTICS

Students should be able to see what they know, what they should know, and what they can do to meet their academic goals based on data analysis. In general, schools amass massive quantities of data on students' attendance, behavior, and achievement, as well as administrative and perception data from polls and focus groups.

Many college and university executives, on the other hand, are unsure how to integrate analytics into their activities and accomplish the desired outcomes and enhancements. What is it that really works? Is it a declaration of intent to invest in new people, technology, or business models? Is it all of the above, or none of the above?

Over the last 5 years, the United States has changed dramatically, and McKinsey's wide analytics experience has aided transitions in both the public and private sectors. Numerous possible difficulties that organizations may encounter while expanding their analytics capabilities emerged from the discussions and interactions, as well as several practical initiatives that education leaders may take to minimize these difficulties.

Any organization has problems when it comes to advanced analytics, but the obstacles in higher education are amplified by sector-specific issues like governance and talent. Leaders in higher education must solve some or all of the most prevalent impediments, not merely preach about the benefit of analytics.

Putting such a high priority on external conformity, many higher education data analytics teams focus their efforts on creating reports to fulfill operational, regulatory, or statutory compliance. The major goal of these teams is to provide university data that will be used by accrediting organizations and other third parties to evaluate each institution.

The primary goal of these groups is to generate university data that accrediting agencies and other third parties may use to assess each institution's performance. Outside of these activities, all requests are viewed as emergencies rather than normal, important chores. Analytics teams have very limited time to enable strategic, data-driven decision-making in this environment.

6.7.1 CHALLENGES

In higher education institutions, analytics teams typically report to the head of an existing function or department, such as the institutional research team or enrollment management. As a result, rather than serving as a centralized resource for all, the analytics function gets intertwined with the department's agenda, with little to no interaction with senior leadership. The impact of analytics is restricted in this environment because analytics insights are not incorporated into the institution's daily decision-making.

In many higher-education institutions, there is a little motive (and considerable reluctance) to exchange data. As a result, universities lack strong data hygiene, that is, clear standards for who may access certain types of data and written regulations for how that data may be shared among departments. Various university departments, for example, may use various datasets to estimate retention rates for different student segments, and when they meet, they frequently dispute over which set of statistics is right.

To make matters worse, many universities are experiencing difficulty linking the multiple legacy data systems that teams utilize in various activities or working groups. Because of the time, it takes to install, train, and acquire buy-in for technological advancements, even with the help of a software platform provider, it will take 2–3 years for institutions to see meaningful returns from their analytics programs. Meanwhile, organizations struggle to create a culture and processes that are based on the benefits of data-driven decision-making.

Due to budgetary and other constraints, higher education institutions may struggle to achieve market pricing for analytical expertise. Attracting and keeping analytical expertise among graduate students and professors should help colleges and universities, but it may be difficult. Furthermore, for higher-education institutions to be effective in pursuing analytics-driven transformation, executives must be proficient in both management and data analytics and be able to solve issues in both areas.

Implementing Best Practices

These challenges may seem daunting, but when senior executives in higher education institutions work to change both activities and mindsets, transformation through analytics is possible.

Five action steps are suggested by the leaders for fostering success.

Create Mandate for Analytics That Extends beyond Compliance

Senior leaders in higher education must make it clear that analytics is a strategic priority. However, if analytics is to reach its full potential, it cannot be considered only as a compliance cost center. Instead, this group would be a source of innovation as well as the institution's economic engine.

As a result, team leaders must articulate the team's overall mission. According to the leaders, the transformation story will focus on how analytics will aid the institution in facilitating the student journey from applicant to alumnus while also providing unrivaled learning, study, and teaching opportunities, as well as promoting a vibrant, financially sustainable institution.

Create a Central Analytics Team That Reports Directly to Executive Leadership

Higher education leaders must allocate the necessary financial and human resources to create a central department or function to oversee and administer the institution's use of analytics, in order to avoid the drawbacks of analytics teams embedded in existing departments or dispersed across multiple functions. This group may be in charge of running a centralized, integrated platform for gathering, analyzing, and modeling datasets and rapidly generating insights.

The analytics team should report to the institution's most senior leaders—in some cases, the provost—in addition to having autonomous status outside of a subfunction or single department. When given more power over choices, analytics leaders have a greater knowledge of the university's challenges and how they affect the institution's broader strategy. Leaders may more quickly discover datasets that might give useful information to university officials—not just in one area, but across the board—and get a head start on proposing viable solutions.

Gain Front-line Support for Analytics and Create a Data-driven Decision-Making Culture

The analytics team must take the lead in creating meaningful analytics communications across the organization to overcome the institution's cultural reluctance to data sharing. Members of the centralized analytics function should communicate officially and routinely with various departments across the university to achieve this goal.

A hub-and-spoke model can be especially effective, in which analysts sit alongside operating unit staffers to promote sharing and directly assist decision-making. These analysts should act as interpreters, assisting working groups in understanding how to use analytics to solve particular issues while also using datasets from other departments.

Of course, having standardized, uniform methods for processing all university data would assist in reliable analysis. Universities that want to foster a culture of data-driven decision-making do not have to wait 2 years for a new data platform to go live. Instead, analysts should identify use cases, which are locations where data already exists and where analysis can be done quickly to yield useful information. Teams will then share success stories and evangelize the benefits of shared data analytics, inspiring others to start their own analytics-driven projects.

Boost Internal Analysis Skills

Contracting out labor in the near term is absolutely appropriate because the skills gap appears to be impeding schools and universities' efforts to change operations through advanced analytics. While outside knowledge can assist address a skills vacuum, it will never be able to totally replace the requirement for in-house ability; the initiative to drive change across the institution must be owned and led within.

In order to do so, institutions will need to change their approaches to talent acquisition and development. They may have to search outside the box to locate experts who are knowledgeable with fundamental analytics technologies (such as cloud computing, data science, machine learning, and statistics), as well as design thinking and operations. Institutions may also need to offer competitive financial incentives and the opportunity to work autonomously on intellectually challenging projects that will have a long-term influence on students and contribute to a bigger purpose to attract new recruits.

Allowing Greatness To Be the Enemy of Good Is a Mistake

It takes time to launch a lucrative analytics program. Institutions may initially lack some sorts of data, and not every evaluation will provide helpful results, but that is no reason to stop trying. Instead, colleges and universities will use a test-and-learn approach, which involves identifying areas with clear issues and good data, conducting evaluations, implementing necessary improvements, collecting input, and iterating as required. These examples may be used to show the value of analytics to other areas of the company, resulting in increased interest and buy-in.

Recognizing Impact from Analytics

Higher education leaders must devote just as much energy to acting on data insights as they do to enabling data analysis. Analytics is a vital facilitator for colleges and universities to address challenging situations, but they must devote just as much attention to permitting data analysis. Implementation will need significant changes in culture, policy, and processes.

When outcomes improve as a result of a university's effective implementation of change—even in a small setting—the rest of the institution notices. This should help to increase the institutional will to push forward and begin working on other areas of the organization that need to be improved.

Some colleges and universities have already surmounted these challenges and are reaping significant benefits from their analytics deployment. Northeastern University, for example, employs a predictive algorithm to determine which candidates are most likely to be the greatest match for the university if admitted.

The analytics team at the firm utilizes a range of data to create projections, including students' high school histories, previous postsecondary enrollments, campus visit activities, and email response rates. According to the analytics team, examining the open rate for emails was particularly useful since it was more predictive of whether students enrolled at Northeastern than what students said or if they visited campus.

Simultaneously, the university looked at data from the National Student Clearinghouse, which records where candidates end up at the end of the application process, and discovered that the schools it had previously deemed primary competitors were not. Instead, it was facing competition from places it had not even considered.

It also discovered that half of its students came from schools where the admissions office had not visited. Northeastern made some changes as a result of the team's overall analysis, including offering combined majors, to appeal to those most likely to enroll once admitted. Furthermore, the management team moved funding from programs that were underutilized to programs and features that were more likely to attract targeted students.

Northeastern University rose from 115 in 2006 to 40 in 2017 in the US News & World Report national university rankings. This is partly owing to these advancements. In another situation, UMUC was investigating the cause of a decline in enrolment in 2013. It spent a lot of money on sales and generated a lot of leads, but the conversion rates were low.

The university's data analysts evaluated the university's returns on investment for its marketing efforts and found a bottleneck: UMUC's call centers were overburdened and understaffed. The university invested in new call-center facilities and saw a 20% increase in new student enrollment in just one year while spending 20% less on advertisements.

The benefits stated are only the tip of the iceberg; advanced analytics' next wave will enable unique, personalized student experiences, with education adapted to individuals' specific learning styles and proficiency levels, among other things.

6.7.2 EDUCATIONAL DATA ANALYTICS TECHNOLOGIES

Data analytics, in general, refers to techniques and tools for analyzing large sets of data from various sources in order to support and increase decision-making. Data analytics, despite the fact that it now encompasses established technologies utilized in real-world environmental, enterprise, and health systems, was only recently examined in the context of Higher Education and School Education.

Data Sources

Educational data is generated by a variety of sources, both internal and external to the school, for example:

- Student information, such as demographics and academic performance in the past.
- Information about teachers, such as their qualifications and work experience.
- Lesson plans, evaluation techniques, and classroom management are examples of data produced during the teaching, learning, and appraisal procedures, both within and outside the physical classroom premises.
- Human resources, infrastructure, and financial plan, which includes educational and noneducational personnel, hardware/software, and budgeting.
- Student well-being, social and emotional development, including support, diversity respect, and special needs.

Analytics types

Educational data analytics systems are now thought to be helpful in overcoming realistic obstacles to long-term, data-driven decision-making in teaching and learning. Data-driven decision-making in schools is described as "the systematic gathering, analysis, review, and interpretation of data to guide practice and policy in educational contexts." Data-driven decision-making has the goal of reporting, evaluating, and improving school procedures and outcomes.

Given the growing demand for data-driven decision-making, data literacy is a critical skill for classroom instructors. It is causally connected to their ability to use educational data for self-evaluation and growth. Given that instructors' data literacy competence is generally limited by a variety of genuine constraints, data analytics solutions are being examined, with teaching analytics, learning analytics, and teaching and learning analytics being the most promising.

Data-driven decision-making in schools is a worldwide movement aimed at promoting School Autonomy by helping schools to fulfill both external regulatory accountability standards and internal continuous self-evaluation and improvement demands.

Data literacy for teachers is an important part of this process because it allows teachers to use data in their decision-making. This is not always a simple undertaking without the assistance of modern technologies. This section discusses educational data analytics systems as a method of supporting schools in making data-driven choices.

Schools are given more leeway in terms of making decisions, such as curriculum design and delivery, human resource management, and infrastructure upkeep and procurement, in this context. This expanded autonomy, however, comes with a greater need to demonstrate (1) that the school is meeting the criteria of external accountability and compliance with (national) regulatory standards, and (2) that the school is engaging in constant self-evaluation and improvement.

To prove these, educational data might be gathered and analyzed.

They may be classified into three categories:

1. *Teaching analytics* refers to tactics and techniques that enable instructional designers (instructional designers and/or educators) to analyze their work in order to reflect on and improve it before providing it to students.

2. *Learning analytics* is described as the "measuring, collecting, analyzing, and reporting of data on learners and their surroundings for the goal of understanding and optimizing learning and the environments in which it occurs."

3. *Teaching and learning analytics,* which combines teaching analytics and learning analytics to assist instructors in their inquiry by allowing them to reflect on their teaching design using evidence from student delivery.

6.7.3 ALGORITHMS AND MODELS

A broad category of big data techniques is available for education mining. Siemens et al. (2014) presented the major category prediction model and inferential model with pattern and structure discovery algorithms, relationship mining, correlation mining, and visualization. The analytics has been applied to different levels of educational content. At the microlevel, clickstreams, logs are taken for consideration. Macrolevel analysis is employed for huge data, that is, institutional level and meso for textual data. The integrated architecture is (Romero and Ventura, 2020) depicted in Figure 6.23.

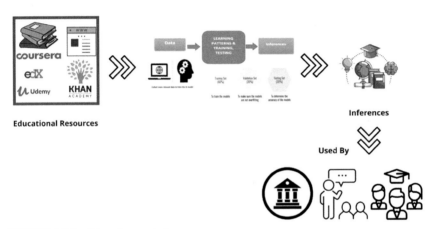

Educational Resources

FIGURE 6.23 Education analytics.

The analytics tools for educational analytics are shown in Figure 6.24.

FIGURE 6.24 Education analytics tools.

The overview of algorithms used in educational analytics is presented in Figure 6.25.

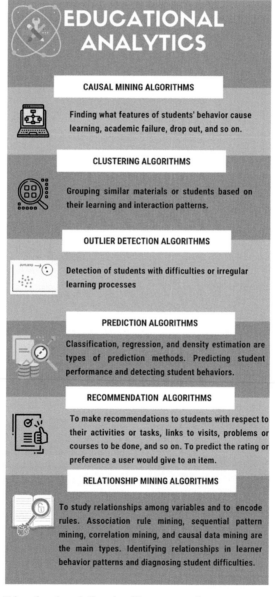

FIGURE 6.25   Educational analytics algorithms—overview.

6.7.4   APPLICATIONS OF DATA ANALYTICS IN EDUCATION

Data Mining in Education

Predictive models are developed using this approach, such as identifying at-risk learners (those who are at risk of dropping out of the course) and assisting teachers in providing intervention to help students succeed.

Curriculum That Is Intelligent and Adaptable

As many curricula as available should be implemented to students using student data. It is possible to create a recommender system based on their preferences (and skills), in which different students may, for example, learn the same content in different ways.

Learning That Adapts

Instructional and learning analytics combines teaching analytics and learning analytics to aid instructors in their research by allowing them to reflect on their teaching design utilizing data from student delivery.

Aiding Management in Making Decisions

In education, data analytics can help with administrative decision-making and resource allocation. For example, they should learn which school facilities the students prefer (or less). It will also provide input to school officials.

Innovation

Models and pedagogical methods can be transformed using learning analytics. The overall goal is to innovate, that is, to pool resources to help students succeed. The purpose of this strategy is not to replace the teacher. It could help professors customize their instruction to maximize learning resources and organize courses in a way that encourages student engagement.

Providing Resources That Are Pertinent to the Learner's Profile and Learning Objectives

It should give students insight into their own learning patterns and make suggestions for improvement.

An alternative to the traditional "end-of-course" assessment

Teachers (with systems) should use data to map students' knowledge domains; it is more than just knowing which question in a specific exam is valid or incorrect; learners' behavior (data and more data) can be analyzed in relation to these maps, that is, we are not interested in whether a student can recall a given definition of a topic; rather, we want to know if he or she can apply the notion correctly.

To summarize, learning analytics focuses on improving current teaching, learning, and, most importantly, evaluation methods. It is not a proposal to abolish existing education; rather, it is a proposal of a teaching machine, in which a machine can be used to educate students. The most intriguing aspect of this forecast is that there was a time when someone believed that education could be achieved without the use of teachers.

KEYWORDS

- **healthcare analytics**
- **image processing**
- **signal processing**
- **genomics**
- **drug discovery**
- **disease prediction**
- **precision medicine**

REFERENCES

Belle, A.; Thiagarajan, R.; Soroushmehr, S. M.; Navidi, F.; Beard, D. A.; Najarian, K. Big Data Analytics in Healthcare. *BioMed Res. Int.* **2015**.

Chang, M-S., Kim, H. J. A Customer Segmentation Scheme Base on Big Data in a Bank. *J. Digital Contents Soc.* **2018**, 85–91.

Gessner, R. C.; Frederick, C. B.; Foster, F. S.; Dayton, P. A. Acoustic Angiography: A New Imaging Modality for Assessing Microvasculature Architecture. *Int. J. Biomed. Imag.* **2013**, *9*, 936593.

https://aabme.asme.org/posts/big-data-in-drug-discovery
https://channels.theinnovationenterprise.com/articles/analytics-and-pricing-decisions
https://data-flair.training/blogs/r-data-science-project-customer-segmentation/
https://doi.org/10.1016/j.inffus.2020.05.009
https://edanalytics.org/
https://en.wikipedia.org/wiki/Learning_analytics
https://ieeexplore.ieee.org/document/6732448
https://indatalabs.com/blog/big-data-behind-recommender-systems?cli_action=161677
2980.296
https://journalofbigdata.springeropen.com/articles/10.1186/s40537-017-0063-x
https://link.springer.com/article/10.1007/s00500-020-04943-3
https://link.springer.com/article/10.1007/s00521-019-04095-y
https://link.springer.com/article/10.1007/s10639-020-10129-z
https://link.springer.com/article/10.1007/s12652-020-02242-1
https://link.springer.com/article/10.1007/s13198-020-00946-3
https://link.springer.com/chapter/10.1007/978-981-15-5421-6_5
https://monkeylearn.com/sentiment-analysis/
https://towardsdatascience.com/fine-grained-sentiment-analysis-in-python-part-1-2697bb111ed4
https://www.byteant.com/blog/7-ways-how-to-use-big-data-in-social-media/
https://www.domo.com/news/press/domo-releases-eighth-annual-data-never-sleeps-infographic
https://www.hindawi.com/journals/complexity/2019/5964068/
https://www.itransition.com/blog/big-data-precision-medicine
https://www.mckinsey.com/business-functions/marketing-and-sales/our-insights/using-big-data-to-make-better-pricing-decisions#
https://www.mckinsey.com/industries/public-and-social-sector/our-insights/how-higher-education-institutions-can-transform-themselves-using-advanced-analytics#
https://www.ncbi.nlm.nih.gov/pmc/articles/PMC5343946/
https://www.ncbi.nlm.nih.gov/pmc/articles/PMC5343946/#B45-ijms-18-00412
https://www.ncbi.nlm.nih.gov/pmc/articles/PMC7327346/
https://www.researchgate.net/publication/261199219_Ontology-Based_Product's_Reputation_Model
https://www.sciencedirect.com/science/article/abs/pii/S0167739X20324912
https://www.sciencedirect.com/science/article/abs/pii/S074756321830414X
https://www.sciencedirect.com/science/article/abs/pii/S1746809418300107
https://www.sciencedirect.com/science/article/abs/pii/S2214579615000155
https://www.sciencedirect.com/science/article/pii/S1532046413001007
https://www.sciencedirect.com/science/article/pii/S2352914818300844
https://www.tandfonline.com/doi/abs/10.1080/10919392.2018.1517481
https://www.tandfonline.com/doi/full/10.1080/17517575.2020.1812005
Lander, E. S.; Waterman, M. S. Genomic Mapping by Fingerprinting Random Clones: A Mathematical Analysis. *Genomics* **1988**, *2*, 231–239.
Masoudi, S.; Harmon, S. A.; Mehralivand, S.; Walker, S. M.; Raviprakash, H.; Bagci, U.; Choyke, P. L.; Turkbey, B. Quick Guide on Radiology Image Pre-Processing for Deep Learning Applications in Prostate Cancer Research. *J. Med. Imag.* **2021**, *8* (1), 010901.

Pardos, Z. A. Big Data in Education and the Models That Love Them. *Curr. Opin. Behav. Sci.* **2017,** *18*, 107–113.

Romero, C.; Ventura, S. Educational Data Mining and Learning Analytics: An Updated Survey. *Wiley Interdiscipl. Rev.: Data Mining Knowl. Discov.* **2020,** *10* (3), 1355.

Zheng, T.; Cao, L.; He, Q.; Jin, G. Full-Range in-Plane Rotation Measurement for Image Recognition with Hybrid Digital-Optical Correlator. *Opt. Eng.* **2014.**

Index